U.S. Marine Corps Physical Readiness Training for Combat

Plus

Marine Corps Martial Arts

Department of Defense

Pentagon
Publishing

UNITED STATES MARINE CORPS
Marine Corps Combat Development Command
Quantico, Virginia 22134-5001

29 January 1988

FLEET MARINE FORCE REFERENCE PUBLICATION............0-1B

MARINE PHYSICAL READINESS TRAINING FOR COMBAT

1. PURPOSE

Fleet Marine Force Reference Publication (FMFRP) 0-1M<u>arine Physical Readiness Training for Combat</u> is intended for use by all Marines. It provides the information and references necessary to establish and conduct physical conditioning programs to prepare Marines for the physical demands of combat.

2. APPLICABILITY

Although not directive, this handbook is intended for use as a reference by all Marines in developing unit physical readiness training for combat. This FMFRP does not establish policy for the conduct of physical fitness training. Refer to Marine Corps Order 6300.3_ for applicable Marine Corps policy on physical fitness. It may also be used as a basis for instruction conducted by unit leaders or as a reference for instruction or training.

3. RECOMMENDATIONS

To improve this manual, the MAGTF Warfighting Center seeks comments and recommendations from the user. Officers and noncommissioned officers with constructive comments are encouraged to respond. Lack of response will be considered as concurrence. The User Suggestion Form which follows can be detached and forwarded to - -

Commanding General
Marine Corps Combat Development Command (WF 06)
Quantico, Virginia 22134-5001

4. DISTRIBUTION

Initial distribution of this FMFRP 0-1B is shown below. Addressees on this list should ensure that appropriate subordinate organizations are provided copies.

JOHN I. HOPKINS
Brigadier General, U.S. Marine Corps
Assistant Chief of Staff, Warfighting
Marine Corps Combat Development Command
Quantico, Virginia

DISTRIBUTION: TDJ

From:

To: Commanding General, Marine Corps Combat Development Command (WF 06), Quantico, Virginia 22134-5001

Subj: FMFRP 0-1B, MARINE PHYSICAL READINESS TRAINING FOR COMBAT, RECOMMENDATIONS CONCERNING

1. In accordance with the Promulgation Page to FMFRP 0-1B, which invites individuals to submit suggestions concerning this FMFRP directly to the above addressee, the following unclassified recommendation is forwarded:

Page________Article/Para No.________Line No.________Figure/Table No.________

Nature of Change:_____Add_____Delete_____Change_____Correct

2. Proposed New Verbatim Text: (Verbatim, double spaced; continue on additional pages as necessary.)

3. Justification/Source: (Need not be double spaced.)

NOTE: Only one recommendation per page.

PREFACE

MARINE PHYSICAL READINESS TRAINING FOR COMBAT

Chapter 1
PHYSICAL READINESS LEADERSHIP
Section I. Role of Physical Fitness in Combat

1101. PURPOSE OF PHYSICAL FITNESS TRAINING

Physical fitness training in the Marine Corps has one purpose: to prepare Marines to physically withstand the rigors of combat. All other goals of physical fitness training are subordinate to and must support attainment of this goal. The idea that only infantry or reconnaissance units and their attachments normally face physically demanding combat is wrong. This error must not influence the priority commanders of combat support, combat service support, aviation, and headquarters units give to physical fitness for combat. Physical fitness for combat has a high priority for all Marines. A sound, effective unit program, requiring limited time and material, offers a greater payoff in combat than many more expensive and time-consuming training programs. This handbook describes the concept and provides the ingredients of a program for commanders to use in carrying out one of their most serious responsibilities: ensuring that their Marines are physically ready for combat.

1102. SCOPE OF THE MANUAL

This handbook provides guidance for all leaders, trainers, and planners of physical training programs. It describes unit physical fitness for combat training in the following chapters:
a. Physical Readiness Leadership. Chapter 1 provides guidance to leaders in the conduct of physical combat readiness training.
b. Physical Readiness Training Programs. Chapter 2 discusses how to structure programs to meet physical readiness goals in a variety of combat training situations.
c. Physical Conditioning Activities. Chapter 3 discusses the primary physical conditioning activities which commanders may use to attain readiness for combat goals. Sections in this chapter address foot marches under load, strength-building activities, and activities which build confidence and the aggressive spirit needed in combat.
d. Combat Water Survival. As an amphibious force, Marines cannot ignore the importance of combat water survival training. Chapter 4 describes a program for commanders to conduct this training.
e. Competitive Conditioning Activities. Chapter 5 describes competitive activities which stress the relationship to combat actions, featuring team-building types of competition.
f. Evaluation of Performance During Training. Chapter 6 describes tests which measure individual and unit physical fitness for combat.
g. The Human Body. Chapter 7 covers the structure and functioning of the body.

1103. COMMANDER'S ROLE

Major General Commandant John A. Lejeune in the 1921 edition of Marine Corps Manual expressed our philosophy of leadership. Among other things, General Lejeune stated that "it will be necessary for officers ... to endeavor to enlist the interest of their men in building up and maintaining their bodies in the finest physical condition." It is significant that the Commandant prefaced these writings in 1921 with the statement that his thoughts were greatly influenced by the experience of World War I. We have never varied as a Corps from this belief that one of a commander's most serious responsibilities is to have Marines physically prepared for combat. Senior commanders and their staffs carry out this role primarily by providing command emphasis, including planning, support, and supervision. Commanders at the battalion/squadron and company/battery level execute a program of physical fitness for combat by close supervision and, most importantly, leadership by example.

1104. HISTORY OF MILITARY PHYSICAL READINESS

Every war has revealed our military physical deficiencies during the initial periods of mobilization. This realization followed the Civil War and has recurred regularly with each national emergency.

a. Training programs in each war were geared to the physical need of the era. Success was dependent upon the amount of time available during training to physically prepare Marines for battle conditions. Frequently, casualties in initial engagements were attributed to the inability of our Marines to physically withstand the rigors of combat over rugged terrain and under unfavorable climatic conditions. With adequate preparation, Marines have always handled the test of battle.

b. During World War II, the first physical conditioning doctrine that could be scientifically justified by testing was introduced. As the war progressed, this program was effective in the physical conditioning of millions of Marines for combat.

c. Postwar periods have traditionally been a time of consolidation. Unfortunately, some leaders considered the conditioning phase of training to be a wartime tool. With such a philosophy prevalent between wars, physical readiness was relegated to secondary importance resulting in a serious lowering of combat effectiveness. The initial commitment of Marines in Korea dramatically displayed this failure to recognize the extreme physical demands of warfare. Call-up of large numbers of reserve Marines and pressing them into the war within weeks after call-up mainly caused the loss in effectiveness.

d. Costly lessons learned from our military experiences over a period of years and the course of several wars led to an increasing interest in the physical conditioning of the individual Marine. No longer can we afford to emphasize physical fitness during wartime and de-emphasize it during peacetime. It is evident that, in spite of increased mechanization and modern weapons, physical readiness retains a vital place in the life of each individual Marine and in every unit within the Marine Corps regardless of mission.

e. Commanders are well aware of the need for rugged and well conditioned Marines. The daily demands of housekeeping, maintenance, support, training, operations, and other time-consuming tasks make it necessary for commanders to set aside time for frequent, regular, and vigorous exercise periods.

1105. PHYSICAL DEMANDS OF COMBAT

There are three primary elements of effective physical fitness for combat: lower body strength and stamina; upper body strength and stamina; and a competitive, combative spirit. Cardiovascular functioning is not treated as a separate element since it is a necessary foundation for the other three. Exercises and activities designed to improve Marines' condition in the three primary elements will foster good cardiovascular and respiratory condition.

a. Lower Body Strength and Stamina. Some common demands that combat places on lower body strength and endurance are marching long distances under load and functioning effectively at the destination; moving quickly and evasively under fire; and carrying wounded Marines to safety.

b. Upper Body Strength and Stamina. Some common demands that combat places on upper body strength and stamina are rapidly emplacing crew-served weapons; handling large-caliber ammunition for extended periods; climbing walls, cliffs, and other high obstacles; and performing field maintenance on air craft or heavy machinery.

c. Competitive, Combative Spirit. While part of this critical aspect of combat readiness is mental and emotional, robust physical condition and a training program which promotes physical aggressiveness greatly enhance a competitive and combative spirit. Activities which force Marines to overcome natural physical fear, which directly leads to fatigue, are particularly valuable.

1106. PHYSICAL EFFECTS OF COMBAT STRESS

It has been thoroughly documented that the added uncertainty and stress of combat have a major physical effect on Marines as well as the acknowledged psychological effect. In other words, in combat, fear equals fatigue. In training, we cannot easily reproduce this combat stress which reduces the effectiveness of individuals and units. However, we can produce fatigue and psychological doubt in training situations by developing a challenging physical training program which stresses the participants. This compels Marines to exercise their ability to continue to function under difficult and trying conditions. Such activities as long foot marches under load and difficult confidence courses train Marines to overcome their own fear and fatigue. Rugged competitive activities such as martial arts and pugil stick fighting prepare Marines psychologically to overcome an opponent as well as their own fear and fatigue. This sort of physically demanding training builds self-confidence and unit morale. The adage that "the unit that sweats more in peacetime will bleed less in battle" certainly applies to its physical training program. Athletic coaches have long followed this approach by such practices as having teams shoot foul shots after wind sprints in order to strengthen their mental conditioning and toughness. Although pushing Marines to their limit is necessary, commanders often set lower standards initially than ultimately desired. The program should then call for steadily increasing the difficulty. A good guide for a starting point is a standard that will be physically and psychologically demanding for no less than three-fourths of the unit. Expressed another way, no more than one-fourth of the unit should be working at less than their limit. The program should progress quickly, augmented, if needed, by remedial training for those unable to meet the unit standards, until the training is challenging for all but the most exceptional Marines in the command. Commanders must never forget that Marines expect to be challenged and thrive on being pushed to their limit.

1107. LIMITS OF PHYSICAL READINESS

A well-conditioned Marine unit has a significant advantage in combat, but this does not mean that physical conditioning can substitute for technical knowledge or good planning. For instance, units must, as discussed in chapter 3, pursue a continuing and challenging program of marches under loads. This does not disprove the axiom that fighting men, if required to carry over one-third of body weight, will generally become ineffective in battle. Leaders and planners must decide what is essential to be carried and must use an effective logistic distribution system, such as unit trains, to make remaining items available when they are needed. In a similar vein, the valid requirement to foster a competitive and combative spirit in Marines does not excuse leaders from the responsibility to practice tactics which will not needlessly risk lives. Physical readiness for combat is a vital goal, but it is only one part of a unit's overall readiness for the lest of combat.

Section II. Fundamentals of Physical Fitness

1201. FITNESS FOR MARINES

Total combat readiness includes technical proficiency and mental, emotional, and physical fitness. If any one of these attributes is lacking, combat effectiveness suffers proportionately. Without technical fitness, Marines lack the knowledge and skill to fight; without mental and emotional fitness, they lack the incentive and will to fight; and without physical fitness, they lack the physical ability and confidence to fight. Physical fitness in a Marine means a whole and healthy body, the capacity for skillful and sustained performance, the ability to recover from exertion rapidly, the desire to complete a designated task, and the confidence to face any eventuality.

1202. COMPONENTS OF PHYSICAL FITNESS

A sound body, free of disease and defect, does not in itself constitute physical fitness. Before an untrained Marine can be considered physically fit for combat, the following traits must be developed:

a. Strength. Every Marine needs enough strength to perform the heaviest task encountered in routine and emergency activities. The basic areas where strength is required are in the arm and shoulder girdle, abdomen, back, and legs. Muscles increase in size, strength, and firmness with regular and strenuous exercises. Without work, they grow flabby and weak.

b. Endurance. Each Marine needs sustaining power to maintain the maximum ability without undue fatigue. There are two types of endurance:

(1) Muscular Endurance. The Marine needs muscular endurance to fight the enemy under the most tiring combat conditions. Muscular endurance is the ability to perform continuous work over long periods of time. Endurance depends on the bloodstream's ability to deliver large amounts of oxygen and nutrition to the muscle masses and then carry away the waste products quickly.

(2) Cardiovascular-Respiratory Endurance. The development of cardiovascular-respiratory endurance ("wind") is necessary to maintain muscular endurance. Cardiovascular-respiratory endurance depends on the efficiency of the blood vessels, lungs, and heart. The maximum effort a Marine can exert over a period of time is limited by the capacity to absorb oxygen and expel carbon dioxide. The average Marine's cardiovascular-respiratory capacity can be greatly increased by exercise.

c. Agility. A Marine must be able to change direction quickly and as faultlessly as possible. The ability to react instantly and to maintain orientation during rapid changes of body position is important to survival. This important ability may be developed by conditioning exercises that require varied and rapid changes of body position on the ground and in the air, such as obstacle courses.

d. Coordination. Coordination is the ability to move all parts of the body in a smooth, efficient, and concerted effort (commonly called timing). A well-coordinated individual does not make useless movements. An individual who moves with precision and accuracy saves energy. Coordination is best developed by practicing diversified muscular activities and skills affecting all body parts.

1203. TYPES OF EXERCISE

Basically, there are two forms of exercise: isotonic and isometric. Both forms are contained in the activities found in this manual.

a. Isotonic. Isotonic exercises are those in which the expenditure of energy is regulated and released during concentrated efforts. The regulated expenditure of energy is controlled by both the mode of exercise and the individual's effort. This type of effort is common to most exercise and sports. To develop endurance, coordination, and flexibility, isotonic exercise must be used. Strength can also be developed or increased through this type of exercise.

b. Isometric. Isometric exercises are those in which maximum effort is applied and held until the engaged muscle energy is depleted during a single contraction effort. The individual exerts full force against an immovable object for a relatively short period of time (6 to 10 seconds) and then repeats the contraction several times with a short period of rest between each contraction. This type of exercise develops only strength; therefore, it has limited application.

1204. PRINCIPLES OF PHYSICAL CONDITIONING

Men and women vary in their physical makeup. Physiological (body) function and reaction also vary in proportion to heavy demands placed upon the body. To attain the maximum benefit without sacrificing Marines' welfare, fitness must grow with a careful program of conditioning. To allow for adjustments in body functioning as the conditioning program progresses and to ensure attainment of objectives, the following principles of physical conditioning must be applied:

a. Overload. As strength and endurance increase, the physical load must be increased until the desired level of fitness is reached.

b. Progression. In beginning stages, the load must be moderate. Gradual progression from this low state of fitness to a higher state is possible through application of a progressive program.

c. Balance. An effective program utilizes various types of activities and provides for the concurrent development of strength, endurance, and coordination as well as basic physical skills.

d. Variety. Some programs fail because the routine becomes boring. The most successful programs always include conditioning activities, competitive events, and military physical skill development.

e. Regularity. There is no easy or occasional way to develop physical fitness. Regularity of exercise is a must, with daily exercise preferred.

1205. THREE STAGES OF PHYSICAL CONDITIONING

Unconditioned or poorly conditioned Marines pass through the following stages in reaching the desired state of physical condition.

a. Toughening Stage. This stage is approximately 2 weeks long and is usually characterized by muscular stiffness and soreness followed by recovery.

b. Slow Improvement Stage. This stage is approximately 6 to 10 weeks long and is characterized by slow and steady improvement until the desired level, or a high level, of fitness is attained.

c. Sustaining Stage. This stage goes on indefinitely in order to maintain the level of conditioning achieved by passage through the previous stages.

1206. EFFECTS OF CLIMATIC CONDITIONS

Temperature, both atmospheric and body, affects the physical performance of Marines. Proper maintenance of body temperature through warm up exercise, proper dress in cold weather, and removal or adjustment of clothing in hot weather is necessary for effective performance and health. Climatic factors to be considered are the following:

a. Exercise In High Temperatures and High Humidity. Marines can endure strenuous physical activity in extremely hot temperatures if they are given an opportunity to become acclimated and if they take enough salt and water. It is essential to continue physical training programs in hot climates. Marines can better withstand high temperatures when they are well-conditioned. High humidity combined with high temperatures presents a serious danger. These conditions prevent the natural cooling of the body by the evaporation of perspiration. Training schedules should conform to the provisions of the base commander in respect to wet bulb conditions. Those who conduct training under these conditions should monitor weight loss and be careful to make fluid replacement adjustments.

b. Exercise at High Altitudes. Certain problems are encountered in conditioning Marines stationed in high altitudes because the heart undergoes greater exertion during exercise. It is particularly important that only light exercise be given initially at such altitudes. A Marine's body gradually adjusts to high altitudes within a few weeks. After this adjustment, progressively greater amounts of exercise are possible.

c. Exercise in Arctic Regions. Military duty in the arctic is so arduous that a high level of physical conditioning is essential. Because of the difficulties of carrying on physical conditioning exercises in extreme cold, Marines should be conditioned to the highest level possible before they arrive. A sustaining program will then maintain that level. When exercising in cold weather, Marines should be required to remove excess clothing to prevent them from becoming damp with perspiration.

1207. WARMING UP AND COOLING OFF

It is a fundamental physiological principle that Marines should warm up gradually before taking strenuous exercise. Such action speeds up the circulation to prepare the body to take an overload and helps to prevent injury to muscles and joints after exercising. Marines should be kept mildly active, walking, stretching, or performing some other mild muscular activity until breathing and temperature have returned to normal. Marines should never be allowed to cool off too rapidly; in cool or cold weather, they should put on additional clothing during the cooling-off period.

1208. PHYSICAL ACTIVITY AS AGE INCREASES

Combat makes severe physical demands on Marines. All Marines, regardless of age, must be physically ready to meet these demands. There is no physiological reason to cease exercise or exertion with age. Increased age usually brings increased responsibility which, in many instances, leads to a routine that can become almost devoid of physical activity. The key to fitness with increased age is to continue exercising at a reasonable level and to include exercise of a vigorous type in the daily routine. Older persons who have not regularly maintained a reasonable state of physical fitness require more time than younger persons to become fit. Such individuals usually require a longer period of time to recover from physical effort than younger Marines. If general health is good, evidence shows that older Marines can develop and maintain a degree of fitness which permits vigorous activity and proper performance of duties. It is both the individual's and the commander's responsibility that all Marines continue a daily sustaining exercise program.

Section III. Goals of Physical Readiness Training

1301. NECESSITY FOR PHYSICAL READINESS TRAINING

A very important objective of training is attainment and maintenance of operational readiness. Marines must be physically ready for operations at any time and under all conditions of climate and environment. A combination of training to develop proficiency in physical skills and conditioning to improve strength and endurance results in physical readiness for combat. The degree of physical fitness required of Marines can be acquired through physical exercises in a normal well-coordinated and closely monitored training routine. The performance of purely military duties, such as drills and marching, is not enough to build all the desired areas of fitness. Few recruits are physically fit for the arduous duties ahead of them. The softening influences of our mechanized civilization and the difficulties of conditioning Marines make physical fitness more important than ever before. If Marines are to be fully developed to and maintained at the desired standard of physical fitness, a well-conceived plan of physical readiness training must be a basic part of every training program. Marines cannot be adequately prepared in any other way for the hard work and arduous demands required on the battlefield.

1302. OBJECTIVE OF PHYSICAL READINESS TRAINING

The overall objective of the physical readiness training program is to develop individuals and units who are physically able and ready to perform their duty assignments or missions during training and in combat. To attain the objective of physical readiness, exercise activities must develop the following areas:

a. Strength and Endurance. Developing adequate strength to perform required duties and adequate endurance to sustain activity over a long period of time.

b. Muscle Tone. Developing muscle tone adequate to maintain proper posture and reasonable weight control.

c. Skills. Developing military physical skills which are essential to personal safety and effective combat performance. As skill is developed, agility and coordination are attained. The essential skills are-

(1) **Marching Under Load.** Marching with individual and unit weapons and equipment.

(2) **Running.** Distance and sprint running on roads and cross-country.

(3) **Jumping.** Broad jumping and vertical jumping downward from a height.

(4) **Dodging.** Changing body direction rapidly while running.

(5) **Climbing and Traversing.** Vertical climbing of ropes, poles, walls, and cargo nets. Traversing horizontal objects such as ropes, pipes, and ladders.

(6) **Crawling.** High crawling and low crawling for speed and stealth.

(7) **Throwing.** Propelling objects, such as grenades, for a distance with accuracy.

(8) **Vaulting.** Surmounting low objects, such as fences and barriers, by use of hand assists.

(9) **Carrying.** Carrying objects and employing man-carries.

(10) **Balancing.** Maintaining proper body balance on narrow walkways and at heights above normal.

(11) **Falling.** Contacting the ground from standing, running, and jumping postures.

(12) Surviving in Water (Or Other Specialized Situations) Using water survival techniques.

d. Character Traits. Jnstilling character traits which help accomplish military missions to include-

(1) Confidence. Confidence develops through achieving progressively more difficult tasks as physical ability grows.

(2) Aggressiveness. Participation in combative activities and contests develops desire and willingness to overcome an opponent.

(3) Reaction Under Pressure. Competitive contests and game situations train Marines to think and to act quickly while under pressure.

(4) Teamwork. Teamwork develops through competitive events in which a number of Marines must coordinate their efforts to accomplish a physical task.

1303. BENEFITS OF EXERCISE

The benefits of exercise are not always understood. Some of the more important results of exercise are listed below:

a. Improved Muscle Tone. Muscular tone improves and, at the same time, muscular strength and endurance are built up.

b. Cardiovascular-respiratory Endurance. Cardiovascular-respiratory endurance, or wind, improves through a process of opening up dormant lung capacity to absorb greater amounts of oxygen.

c. Circulation. Circulation of the blood speeds up and extends to a greater portion of the body as exercise forces the blood to service all parts of the body. The efficiency and effectiveness of the heart, lungs, and blood vessels improve.

d. Flexibility. A wider range of muscular movement is possible and rapidity in physical skills grows.

e. Elimination of Body Waste. Bending and twisting the body and the general speedup of body processes caused by exercise regulate and help eliminate body wastes.

f. Tension. Working off excess nervous energy and relief from daily worries and cares relieve tension. Participation in exercises leaves little time for worry.

g. Sleep. Sleep improves because muscles are healthfully tired after about of exercise. A by-product of sound sleep is relief of tension.

h. Obesity Control.. Control of obesity (fat) is made possible by using up excessive amounts of fat-producing food elements.

i. Injury Susceptibility. Susceptibility to injury is reduced through exercise. Muscles, tendons, and joints are strengthened. Injuries such as hernia, back strain, and joint sprains are less likely to occur if muscles are maintained in proper tone.

Section IV. Leadership Roles

1401. PSYCHOLOGICAL LEADERSHIP

The full development of a Marine's resources is not all physical. To be effective in developing physical readiness, leaders must realize that mind and attitude are also important to success. The more important psychological considerations are to—

a. Promote Understanding of the Value of Physical Readiness. A desire to be physically ready should be created in all Marines. Motivation is increased and Marines take greater interest in their individual physical fitness if they understand the value and benefits of vigorous exercise. When Marines realize their efforts are an investment in their own personal welfare, it should not be difficult to obtain their cooperation. Marines should understand the objectives, the benefits, and the value of each type of exercise activity in their program. They should also understand the relation of physical readiness to survival in combat.

b. Maintain a Positive Approach. Physical readiness training for combat is strenuous and demanding. It is a responsibility of leadership to create an atmosphere where all desire to participate fully. This attitude should be fostered. A negative approach must not be identified with physical readiness training, even with those having difficulty. Only in unusual cases should fear of punishment be the motivating factor behind good performance. For those few who cannot keep up or attempt to malinger, an effective remedial program is essential.

c. Seek Cooperation and Develop Morale. In a program placing maximum physical stress upon individuals, it is necessary to gain their cooperation. Favorable reaction is enhanced by proper planning and organization, challenging requirements, use of competition, and application of a progressive program resulting in physical fitness. As physical fitness grows, morale also grows.

1402. COMMAND AND SUPERVISORY FUNCTIONS

a. Command Functions. Commanders should take the following actions to support physical readiness training:

(1) Lead by personal example.

(2) Instill command interest and indicate to subordinate personnel the importance of this training to the welfare of the organization.

(3) Allot sufficient time for the achievement of objectives and monitor the use of such allotted time. The substitution of other training or routine duties for scheduled physical readiness training is unsound and unwise.

(4) Assign and properly utilize qualified personnel to supervise and conduct physical readiness training. If leaders are not competent, take action to ensure they become competent quickly. Reassign those who do not meet this standard to other duties.

(5) Make necessary facilities and funds available to support a program to develop physical readiness within all personnel.

(6) Measure the physical fitness of individuals and units in order to evaluate progress and to determine if the program is successful.

b. Supervisory Functions. Leaders responsible for planning, conducting, and supervising physical readiness training should take the following actions:

(1) Prepare physical readiness training schedules which apply the principles of physical conditioning and which aim for a particular type of program plan.

(2) Provide for wide participation of as many Marines as possible. All Marines, regardless of position or age, will benefit from regular exercise. In some instances, special efforts are necessary to overcome obstacles to regular and frequent training. Special effort is also necessary to ensure remedial conditioning. Such conditioning should occur for those who are physically substandard and after extended absence due to leave, sickness, injury, and travel.

(3) Prevent waste or unwise use of time allotted for physical readiness training. Time-wasters include unprepared instructors; assignment of one instructor to a group larger than a platoon; progression which does not keep pace with the physical development of the Marines; extreme formality; inadequate equipment or facilities which require waiting turns to exercise; and lengthy rest periods between exercises which interfere with the application of overload.

(4) Ensure that the program contains vigorous physical activity. Such activity places progressively greater demands upon the body during each exercise session and also over the duration of the training program. To be of benefit, exercise must tire the muscles and cause the heart to increase its rate of beat.

(5) Set an overall objective for each physical fitness program.Observe the training as necessary to ensure that the established objectives are being achieved.

(6) Observe physical readiness training to insure the use of a positive approach. To implement a positive attitude, small-unit leaders and instructors should personally set the example; have an understanding, fair, and sympathetic attitude; recognize individual differences; and motivate Marines toward their best effort.

(7) Guide and inform small-unit leaders and instructors concerning approved techniques, directives, and literature. As necessary, arrange for local training of instructors to include clinics, conferences, schools, and demonstrations.

(8) Determine the effectiveness of physical readiness training by personal participation in and observation of training, analysis of field inspection reports, and analysis of individual physical fitness test scores. Scores may be combined to reflect the fitness of the unit.

1403. SMALL-UNIT LEADERS AND INSTRUCTORS

a. Responsibility. The instruction and conduct of physical readiness activities are the function of company/battery commanders, platoonleaders and persons assigned as instructors. Experience has proved the effectiveness of physical fitness development when conducted in company-and platoon-size units under direct control of the leader with overall supervision by the parent-unit commander. For example, all the platoons of a company may exercise at the same time under the general supervision of the company commander, with each platoon conducting the assignment separately and under its own leadership.

b. Leader's Assignment. A small-unit leader or an instructor in a school or training activity is assigned to a combat unit or to a support unit. In this assignment, the leader is responsible for all training to include physical readiness training. In a different situation, the noncommissioned officer (NCO) or officer may beassigned as a full-time physical readiness instructor. This contrasts to a unit leader assignment where only part of the time is devoted to such training. In either case, the leader will hold an important and vital position for the physical fitness of Marines.

c. Leader's Training. Leaders may come to the assignment either fully or partially trained. It may be their first responsibility for the development of physical fitness. If they have had previous training through experience, make certain that their information is supplemented with study of this manual. If they have had professional training in physical education during civilian life, but no military experience, they should also use this manual to learn the methods used by the Marine Corps. A new leader should take advantage of various ways to learn including attendance at leader training courses, self-study, practice, and discussion with more experienced leaders.

d.Leader's Objective. As a physical readiness training instructor, the leader has two general objectives. The first is to motivate Marines to want to be physically fit. The second is to conduct a program that will develop a high degree of physical fitnss. Motivated Marines will react enthusiastically to such a program. It aids greatly in achieving local program objectives.

e. Leader's Personal Fitnes. A unit leader who must instruct and demonstrate physical activities must be in physical condition to do the job without undue physical stress.The leader should be able to do those things that must be demonstrated. The leader's strength, endurance, posture, and skill should set the example. This does not mean that the leader must excel, as other Marines do not expect championship performance. However, they do expect, and deserve, a creditable showing of fitness for the job.

f. Leader's Knowledge. The leader must have three types of knowledge to properly administer physical readiness training. Theyare- -

(1) Knowledge of Marines. The leader must understand Marines, know how to lead and motivate them, understand how they learn, and apply this knowledge wisely in the day-to-day training situation.

(2) Understanding of Body Functioning. A more intelligent exercise program results from understanding and applying the principles which govern physical conditioning of the body. The leader with such knowledge can better prescribe, adjust, and regulate exercise types, amounts, and progression to attain fitness.

(3) Understanding Exercise Activities. The leader needs to understand the contribution each type of physical activity makes to physical fitness, and how to use each activity to develop fitness. Skill to demonstrate and lead the various activities is a necessary part of technique and is invaluable to the instructor or small-unit leader.

Chapter 2
PHYSICAL READINESS TRAINING PROGRAMS
Section I. Development of a Program

2101.GUIDANCE FOR PLANNERS

This chapter instructs planners on physical readiness training procedures. It contains program planning guidance including factors to consider when developing programs; steps in assembling a program; definitions of activity packages and systems of exercise; selection of activity packages; and selection of systems used in implementing those packages.

2102. MARINE CORPS MISSION

The mission of the Marine Corps is to seize and defend advanced naval bases and to perform other missions as the President of the United States may direct. Traditionally, to perform this mission, Marines have been projected into the area of operations by Navy vessels and have attacked hostile beaches over the shore in landing craft, assault amphibian vehicles, and helicopters. Recently, performance of this task has been somewhat complicated by the introduction of the concepts of maritime prepositioning and the air-landed Marine expeditionary-brigade. Under these concepts, Marines, landed at airfields in foreign countries, will join their equipment at commercial ports. The distance between these airfields and ports is often 10 or more miles. In time of international crisis, it can be expected that transportation assets will not be adequate. Marines from all elements of a Marine Air-Ground Task Force may have to march from the airfield to the site of their equipment. All Marines must be physically conditioned to rigors of conducting foot marches with individual weapons and equipment.

2103. EVALUATION OF FITNESS

The Marine Corps mission is such that all Marines must constantly and consistently achieve a high level of fitness that prepares them for the demands of combat. While Marines have many different MOS's, all Marines must be prepared for the demands of marching under load and performing basic infantry tasks, such as rear security and patrolling. While the physical fitness test remains our universal measure of individual fitness, it is the commander's responsibility to observe and evaluate the unit's ability to perform effectively in combat. Often the commander's most effective evaluation tool is weekly physical training sessions. The commander should participate in these sessions and see that they are fully integrated into the unit physical conditioning program.

2104. INTEGRATION OF TRAINING

Training time must be used efficiently and wisely. Every opportunity to integrate physical training into other training activities should be seized. For example, foot marches under load can be integrated into the normal daily activities as a way to move from one training or work site to another. This integration reflects realism in training and should be used to maintain overall proficiency. To accomplish this requires imagination and enthusiasm on the part of the commander. The area of operations in which the training is conducted determines what can be integrated and how. If possible, training should also consider the terrain and climate of the area in which the unit will subsequently conduct training or operations. Training should include familiarization with special equipment and the application of specialized techniques to tactical principles. Activities such as gun drills and command post emplacement drills also have physical conditioning value, particularly if supervised and done competitively or measured against time standards.

2105. ACTIVITIES AT THE UNIT LEVEL

Physical readiness training is a command responsibility and is generally conducted in Marine units at the company and platoon levels. Battalion/squadron commanders must constantly supervise, making sure that all Marines are physically prepared for combat.

a. Physical readiness training should usually be conducted weekly by each battalion/squadron-level unit, using events designed for all personnel (e.g., marches under load, battalion/squadron runs, military field days, etc.).These events enable the commanders to demonstrate personal leadership, and observe and evaluate the physical condition and combat physical readiness of their Marines. This exercise period should be designed to elevate the unit's morale and emphasize unit identity through the wearing of uniforms and the carrying of battalion and company guidons. The battalion/squadron commander should make every effort to be a visible participant in the unit's physical readiness program on a weekly basis. These activities could be conditioning marches under load or a weekly calisthenics/unit run.The suggested times for these activities are early on Monday morning or late on Friday afternoon after the training week has been completed. The uniform may be physical training gear orutilities and boots.

b. Commanders should be cautious of long runs in boots and utilities. This is not to suggest that units should not run in utilities, but to remind commanders that boots are designed for marching, not running. If commanders choose boots and utilities, they should select off-road routes which will provide a cushioning effect.

c. Units should be made well aware of the commander's intent before the unit falls out for exercise. If the exercise period is conducted early in the morning, it need not necessitate an extensive shower/clean-up period afterward, so long as all Marines from the commander on down remain dirty. Commanders should inform their Marines not to fall out for these sessions in clean, pressed uniforms.

d. The physical readiness training of headquarters units is often difficult to manage. Often physical training is left for individuals or sections to develop and manage on their own. This approach can be effective, but if not supervised, it can result in a "paper" program. Headquarters unit commanders should endeavor to conduct weekly integrated training which will enable them to evaluate the physical readiness of their personnel. Often this training has to be scheduled before or after normal working hours to maximize participation of the entire unit.

2106. STEPS IN PLANNING

To implement workable and effective programs (as directed by CMC ALMAR 261/87) the planner must-

a. Determine the Type of Program Needed. Marine units are inherently different in organization and mission. The physical readiness program must be tailored to the mission and to the current physical condition of most unit personnel. Programs to meet this need are of the following types:

(1) **Developmental Programs.** Marines in a beginning or poor state of physical readiness need a program which will develop strength, endurance, physical skills, and character traits which are beneficial to successful accomplishment of military missions. Such programs should be applied progressively to rise gradually to a peak of fitness and skill.

(2) Maintenance Programs. Once Marines reach the sustaining stage of conditioning, their goal is then to maintain this level by participation in a maintenance program.

(3) Remedial Programs. The term "remedial" is usually applied to those individuals or groups who possess substandard physical fitness. For example, a remedial physical conditioning program could be applied to persons who are overweight, who fail to reach physical fitness test standards, or who have missed extended periods of conditioning due to illness, injury, extended hospitalization, or other absence. Saturday training sessions are designed as remedial sessions for all individuals who either miss the daily training or have failed to meet adequate standards. Note that personnel in light duty or no-duty status should be expectedto attend all training sessions, observing, supervising, or assisting as needed.

b. Determine the Time Required. The amount of time for training operations varies considerably. However, every unit can find time to conduct physical readiness training. Frequent (i.e.,daily) physical training of short duration-30 to 60 minutes-is preferred over occasional longer periods. Other demands for training time are so urgent that every minute of time allotted for physical readinesstraining should be used. Determine the time required per week anddivide it into daily blocks.

c. Organize for Various Group Size. It is essential to stress exercise rather than formality. Marines must complete the program where they are- -on the training field, in the motor pool, on the range, next to the classroom, in the office area, in the shop, aboard ship, or elsewhere. Although desirable, it is not always possible to assemble company-size units for physical training. Platoon-size groups are appropriate for the proper conduct of physical conditioning activities. Certain situations may require exercise programs for section- or squad-size units. This manual outlines programs for all situations and types of organizations.

d. Allow for Weather and Exercise Area. In programing and scheduling, the climate and terrain often govern the selection of activities.

(1) Weather changes cause differences in temperature, rainfall, wind chill, and snow. These changes should be anticipated as they dictate the type of program. Alternate plans should be part of the schedule. Nevertheless, normal weather changes and rain should not drive training inside.
(2) Local terrain and available exercise areas may also influence the selection of activities and the type of program which it is possible to support. Some activities can be completed in nearly any area.
e. Plan for Seasonal Change. As most physical readiness training is conducted outdoors, it is necessary to recognize seasonal change. A program should be divided into fall, winter, spring, and summer parts. Seasonal change also causes change in light. For example, an early morning program started in the summer will have ideal lightconditions, yet in the fall or winter, darkness will occur at that same hour and interfere with the conduct of the program. Develop programs in seasonal blocks, and make provisions for anticipated changes.

f. Consider Needed Facilities. An excellent program can be conducted with practically no facilities since there are exercises which require no equipment. However, a better program can be developed when supported by certain facilities and items of equipment. Proper command support, plus ingenuity, will solve this problem. (Items of equipment, when necessary to support the recommended exercises, are included in the chapters on exercise activities.)

g. Specify Appropriate Uniform. The uniform worn for exercising depends upon the season of the year, the state of the weather, and local regulations. Whenever possible, Marines should be dressed alike. Undershirts are preferred as the upper garment when the weather permits. A uniform that restricts the free movement of the body should not be worn when exercising.

h. Consider Availability of Instructors. Leaders who can lead and direct the scheduled activity must be available. Organizational units should train junior officers and noncommissioned officers down to squad or section leaders to instruct and lead the various activities.

i. Select Activity and System. With the type of program needed and the objectives in mind, select an activity package or an integrated training package for each day's scheduled physical training. At this same time, the system to be used in employment of the selected activity or activities must be determined. This selection affects equipment, areas, instructors, transportation, and other support requirements.

j. Secure Command Participation and Support. Prepare and brief the commander to assure full understanding of the objectives and administration of the program. The full participation and support of the commander will greatly improve the level of success.

k. Supervise Execution. Determine needs, publish the program, and supervise its execution as a necessary part of developing the unit's training schedule.

2107. SELECTION OF ACTIVITIES AND SYSTEMS

a. Exercise Activities. The planner must determine exercise activities which will be appropriate to include in the program. An exercise activity is a single means of exercise usually identified by the name applied; for example, running, log exercises, and obstacle course. Many exercise activi ties can become part of activity packages.

b. Activity Packages. Many of the physical activities described in this manual are arranged in prescribed sequences and are known as activity packages. An activity package is a number of exercises of the same type, assembled as a group or a set, and arranged in a specific sequence. Exercise packages are organized in such manner that not more than 15 minutes will be required to complete the execution of any package. Each type of activity is explained in laterchapters. The number of available drills, tables, or circuits; the manner of organization; and the contribution each makes to the total program are covered. Full understanding of this information will greatly assist in developing effective programs. Various designations are used to identify exercise packages; for example, conditioning exercises when arranged in a set order are known as drills, and other packaged activities are designated as tables or circuits. The following are activity packages:

(1) Activity packages for groups:

o Conditioning Drill Two
o Conditioning Drill Three
o Rifle Drill
o Log Drill
o Grass Drill
o Running Tables
o Guerrilla Tables
o Circuit-Interval Table
o Combatives Tables
o Relay Tables

(2) Strength Circuits:

o Fixed Circuit
o Movable Circuit
o Simplified Circuit (Circuit Interval Table)

(3) Activity packages for individuals:

o The 6-12 Plan
o Weight Training
o Isometric Exercise

c. Advantages of Using Packages. These of exercise packages simplifies scheduling and conducting of exercise and results in the following benefits:

(1) Schedule development is simplified as the planner assembles packages which will satisfy the training objective. There is no need to deal with selection of individual activities or to be concerned about the amount of time to be expended on each.

(2) Any 15-minute period, and in some cases less time, can be scheduled or used to perform an activity.

(3) If longer periods of time are available or if the objective demands, several packages can be assembled to provide a more complete period of activity.

(4) Marines are assured a balanced set of exercises or activities as each package is carefully arranged to reach all muscle groups.

(5) The instructor can concentrate on the conduct of a vigorous workout as the type and duration of the activity have already been determined.

d. Nonpackaged Activities. Several types of activity are not packaged. Activities in this category are conditioning marches, unit runs, obstacle courses, combat water survival swimming, teamcontests, and team sports. These activities can be scheduled in combination with packaged activities, or they may be scheduled separately. Often they require a longer period of time. Most nonpackaged activities require a 50-minute period to satisfactorily complete their objective. For conditioning marches, a few hours are required. The benefits of these activities should not be overlooked as some desirable objectives cannot be attained without them.

e. Systems of Exercise. Several methods or systems of organizing exercise and activity packages can be used. Each system is based upon a specific organization as follows:

(1) Single Activity System. The unit leader immediately assumes command of the unit at the beginning of the exercise period. The leader moves the unit to a predesignated exercise site at double time, forms the unit in a circle around the leader, groundsclothing and equipment as appropriate, and quickly moves into the exercise routine. There is usually no time to teach; therefore,the Marines must know the activity to be used. At the conclusion of a 5- to 15-minute period, the unit leader returns the unit to the instructional area at double time and releases the unit for the next scheduled activity.

(2) Progressive Activity System. All Marines (company or platoon) complete activities in the same order during the period. For example, Drill One is followed by dual combatives, and finally a 1-mile run. This system is usually progressive from a warm up activity, such as Drill One, to an activity which contributes in a major way to one of the objectives, such as aggressiveness development through combatives or cardiovascular development, such as running.

(3) Rotating Activity System. Set up the same number of activities or stations as there are platoons in the company. Each platoon rotates through each station in turn. With four platoons in a 50-minute period, about 10 minutes can be devoted to each station. With three platoons, approximately 15 minutes may be spent at each station. Activities must be of a type that can be covered in the time allotted. For example, with three platoons, Station 1 could be Conditioning Drill One; Station 2, running; and Station 3, a team contest.

(4) Circuit System. Set up a number of stations to provide various types of exercise equipment and/or items of apparatus. The idea is to keep all Marines busy and exercising vigorously for a short period of time at each station. The fixed strength circuits and the movable strength circuits are examples of this system. Station changes must be rapid, and the exercise must be started quickly after each change. Since the objective is to exercise at top speed, the motivation comes from frequent change of activity by moving to another station. Rotation by station groups continues until all Marines have covered all stations.

(5) Interval System. This system stresses the development of strength and endurance. It involves heavy work for a given distance within a specified time, alternated with lighter work and recovery, but never stopping during the workout. This procedure is repeated, and the intensity is increased gradually as physical condition improves, but always with adequate recovery. The important concept is stress, recover, stress, recover, and so on. This system is often applied through running, but other activities of a continuous nature may also be used. An example is the Circuit-Interval Table.

f. Steps in Selecting Activities and Systems. The planner should follow these steps.

(1) Evaluate the needs of one unit with emphasis upon unit mission, objective, and time available.

(2) For each day, select an activity package or a combination of activities which will contribute to the objective.

(3) Then determine the system to be used in implementing the selected activities. There are several possible choices and much opportunity for flexibility in program development. For example, the weekly program may contain various exercise packages, systems, and time periods. Additional guidance is contained in following chapters.

2108. SAMPLE WEEKLY PROGRAMS

The variety of weekly programs given shows available alternatives. Programs and schedules other than those illustrated may be assembled to provide schedules to fit any situation. Many other scheduling combinations are possible through use of the packaged activities, varied time periods, and different types of assigned duty.

a. Sample A. Figure 2-1 illustrates a 60-minute time allotment in which the single activity system is used with activity packages that vary from day to day. In this program, 5 hours of training are scheduled for the week. This program needs some type of integrated unit training to supplement it each week. A bimonthly conditioning march of 10 miles in 3 hours is suggested to round out this program. The Saturday session is designed for remedial physical training.

NOTE: A physical training activity must be conducted at a minimum of once every 72 hours. After 72 hours, muscles begin to atrophy.

Figure 2-1 is available in the illustration section at the end of the book

b. Sample B. A more comprehensive weekly program (fig. 2-2) illustrates the scheduling of various-length periods, a variety of activities, and the use of three systems during the week. In this schedule, 3.5 hours are included for the week. The following should be noted concerning this schedule:

Figure 2-2 is available in the illustration section at the end of the book

(1) On Monday, a 60-minute period is available for physical readiness training. Three 20-minute packages are scheduled. Each platoon, under its own leadership, will progress through each scheduled activity in turn.

(2) On Tuesday, Wednesday, and Thursday, only 30-minute periods are available. Here each platoon leader supervises each platoon in a single activity system.

(3) On Friday, the company commander directs training to exhibit personal leadership example and supervision. In this example, only one strength circuit and one cross-country course are available, and each will accommodate only one platoon. For this reason, Conditioning Drill One station is included. A platoon rotates to each station within the 60-minute period.

(4) The Saturday session once again is designed for remedial physical training.
(5) Bimonthly, a 10-mile/3-hour march under load will be necessary to round out this program.
NOTE: Cardiovascular/respiratory exercise to improve endurance must be of long enough duration (20 minutes or more) to induce and maintain oxygen debt.

c. Sample C. A third sample schedule (fig. 2-3) illustrates the use of the single activity system for 5.25 hours of training. This schedule includes sustaining-type activities for a unit that has passed through both the toughening and slow improvement stages of conditioning. The use of competitive activities is featured to holdinterest and provide self-motivation. A bimonthly 10-mile/3-hour march under load will be necessary to round out this program. The Saturday session once again is designed forremedial physical training.

Figure 2-3 is available in the illustration section at the end of the book

NOTE: Thc body will respond to hard work or physical training without soreness and stiffness provided the work or physical training is conducted regularly.
NOTE: Excrciscs of short duration increase strength and bulk but do little to effect endurance.
NOTE: Thc times shown in the charts arc participation times.

Section II. Remedial Physical Training

2201. NEED FOR ATTENTION

Remedial physical training is a process by which physically substandard individuals are conditioned to meet standard levels of performance. To achieve a full degree of operational readiness, it is necessary to bring all Marines up to the prescribed standard. Experience has demonstrated that some Marines have difficulty due to a poor state of fitness, obesity, or lack of motivation. Attention to these individual deficiencies will help improve unit combat readiness.

2202. IDENTIFICATION OF INDIVIDUALS

The company or battery commander identifies persons who cannot achieve the prescribed level of proficiency through the physical fitness test or as a result of observation during training. The commander notes particular weakness of body parts as indicated by failure of certain test events. These Marines are then placed in a special remedial program either at platoon, company/battery, or battalion/squadron level. Extra time is devoted to overcoming the weakness. These extra periods of conditioning may be during or after duty hours, as determined by the commanders. Other persons need reconditioning as result of hospitalization. Restoring physical fitness to damaged areas of the body is accomplished through progressively graded physical activities under professional supervision, not by the physical fitness trainers.

a. Types of Deficiencies. Physical deficiencies which can be corrected by exercise fall into several categories.

(1) Lack of Strength in One or More Body Parts. The major muscle areas concerned are the arms and shoulder girdle, back, abdomen, and legs.

(2) Lack of Overall Endurance. Usually, there is deficient muscular and cardiovascular endurance.

(3) Deficiency in Coordination and Agility. In these cases, physical skill is not developed to a satisfactory degree in activities such as crawling, running, jumping, climbing, traversing, vaulting, pushing, pulling, lifting, and carrying.

(4) Overweight or Underweight. Either condition may interfere with physical fitness and mission accomplishment. Lack of exercise is not always the cause. The cause may be malfunction of normal physiological functions or it may be poor health habits such as over- or under-eating, lack of adequate rest, or overconsumption of alcohol.

(5) Lack of Motivation. Not all persons are motivated to attain or maintain **a** desirable state of fitness. Some Marines do not understand the importance of physical fitness, some find proper exercise too difficult, and others find it inconvenient.

b. Causes of Deficiencies. There are several causes for Marines being deficient in physical fitness:

(1) Absence of exercise.

(2) Exercise which fails to develop all muscle groups and components of fitness.

(3) Exercise which is not vigorous enough or which lacks progression.
(4) Injury or illness which depletes fitness.
(5) Inadequate amounts of sleep or rest.

c. Methods of Detection. The commander has several means by which to detect physical deficiencies:

(1)

from those who pass. Make an analysis to determine the cause of failure as revealed by test scores.

(2) Observe Marines as they perform physical tasks-

o Marines who have difficulty during training or in physical types of work,

o Marines who have become obese and therefore experience difficulty.

(3) Be alert to those Marines who are often on sick call, returnees from hospitalization, or newly assigned.

2203. ADMINISTRATION OF REMEDIAL ACTION

a. Group. Attitude Marines who are singled out as being physically deficient are self-conscious and are not always convinced they need extra help. Within a deficient group, motivation may be low. These Marines must be convinced that a remedial program which is tailored to their needs will help them and will eliminate their deficiencies. Encouragement is often needed and desirable.

b. Leadership. The leader of this group must study individuals and know their deficiencies. The unit leader should counsel Marines individually, observe them closely as they progress through the remedial program, maintain records, and adjust the program as required.

c. Measurement.. Whenever improvement in performance is noted, Marines should be measured by the physical fitness test either individually or by group. When testing reveals an individual to be satisfactory, that individual should be released from the remedial group. There may be exceptions to this policy in the case of Marines who are overweight or in the execution of an exercise program prescribed for some special purpose.

d. Organization of a Remedial Group

(1) A remedial group is usually a small group of Marines within a company or battalion. In some situations, it may be a larger group numbering 50 to 75 Marines. In the case of a larger group, divide the Marines initially into subgroups according to ability. Prescribe exercise loads commensurate with their ability. General conditioning is sufficient in the early part of the program to qualify Marines who are on the borderline. Qualification will reduce the size of the group, permitting more individual and specialized attention to those who remain.

(2) As the program progresses, regroup individuals who have like deficiencies in order that they can concentrate on their weakness. For example—

(a) One group may be weak in the arms and shoulders as revealed by failure in the chin up event. This group, with an assistant instruction charge, can work on pull-ups, rope climb, pushups, rifle or logexercises, horizontal ladder, and similar types of exercise.

(b) Another group may be weak in general endurance as revealed by the3-mile run. This group could profit from participation in conditioning drills, running, grass drill, and strength circuit.

(3) It may be necessary to form some groups to overcome weaknesses in skills, such as an inability to throw, to quickly change direction while running, to crawl rapidly, or to carry a load. Lack of coordination or lack of practice may cause these deficiencies. In this instance, instructors must provide an opportunity to practice and correct poor form and other errors as they are noted.

Chapter 3
PHYSICAL CONDITIONING ACTIVITIES
Section I. Marching Under Load

3101. GENERAL

Few physical fitness activities are as directly related to readiness for combat as foot marches under load. In addition to the obvious physical conditioning and unit cohesiveness benefits, marches under load prepare most Marines for numerous foreseeable tasks in combat. The idea that only infantry and reconnaissance units actually require training marches to prepare them for combat is clearly a mistake. In addition to the very real possibility of Marines from all types of units being used to fill combat-depleted infantry units, it is likely in maritime prepositioning force operations that aviation, combat service support, and command element personnel will have to march with weapons and equipment, from the arrival airfield to the equipment marshalling site. In American coastal cities, the harbor is nearly always 10 or more miles from the airport. It is difficult to imagine any different condition in a lesser developed country .A further combat-related benefit is that foot marching under load is the most effective way to develop leg strength in the context of a unit training program. In his classic study of troop performance in World War II, Men Against Fire, S.L.A. Marshall noted that the greatest single weakness of replacements of all occupational fields was leg strength. Truck drivers must push trucks which are stuck in mud, and headquarters communicators must climb hills to emplace antennas. The demands on the lower body in combat apply to all Marines, and foot marching is one activity which can help build these muscles while being conveniently integrated with the rest of the training program.

3102.TRAINING GOALS

The standard for success of a foot march is very simple to measure: did the unit arrive at the destination at the prescribed time with Marines in condition and required equipment present to accomplish the mission?A progressive program can increase a unit's readiness. It can instill pride by increasing distances and rates of march and by selecting routes over increasingly challenging terrain as the Marines become better conditioned. However, the ability to execute the mission at the conclusion of the march must remain the standard of success.

3103.MOVEMENT PLANNING

The basic considerations in planning a foot march are the mission, tactical situation, terrain and weather, and the units to participate. The success of the march will depend largely upon the thoroughness with which it is planned. A successful march is characterized by adherence to prescribed routes and time schedules, the efficient employment of the means available, and the ability of the unit to accomplish its assigned mission upon arrival at the destination.

a. Movement Orders. Movement planning culminates in the preparation and issuance of an operation order prepared in the standard, five-paragraph format. Necessary annexes are attached to furnish detailed information required for the movement. Written movement orders are rarely prepared at company level. A discussion of the annexes is contained in this chapter.

b.March Planning. March planning, as discussed here, is the planning conducted at battalion level. March planning may be organized into the following steps:

(1) **Preparation and Issuance of the Warning Order.** In order to afford subordinate units the maximum possible time to prepare for a pending move, a warning order containing all available information about the march is issued. The amount of planning time available will determine the time of issuance and the content of the warning order.

(2) Estimate of the Situation. In his estimate, the commander considers the mission, terrain, weather, time, and space factors, available routes, available transportation for the movement of equipment and/or shuttling of Marines, enemy capabilities, disposition of own forces, physical condition and training of Marines, and courses of action available to the command.

(3) Organization and Dispatch of a Reconnaissance. Party very march plan is based on as thorough and complete a ground reconnaissance as time and the situation will permit. Map and aerial reconnaissance are valuable in formulating a plan, but are not a substitute for ground reconnaissance. Route reconnaissance is accomplished by a reconnaissance party which usually consists of a reconnaissance element, an engineer element from the attached or supporting engineer unit, and a traffic control element. Unit standing operating procedures generally establish the basic composition of the reconnaissance party. It is modified as necessary to meet the requirements of a particular march. A recommended method of reporting information obtained by the route reconnaissance party is shown in figure 3-1.The minimum information required from the reconnaissance party is—

(a) Available routes and conditions.(Routes may be specified by higher headquarters.)

(b) Recommended rate of march.

(c) Selection of start point and release point, or confirmation of the suitability of start point and release point previously selected by map reconnaissance.

(d) Confirmation of location of the assembly or bivouac area.

(e) Location of critical points on the route.

(f) Distance between critical points on the route and total distance from start point to release point.

Figure 3-1 is available in the illustration section at the end of the book

(g) Location of obstacles and estimation of necessary Marines and equip- ment needed to repair and maintain routes.

(h) Number of guides required and their location on the route.

(4) Development of Detailed Movement Plan. Some of the elements included in a detailed movement plan are-

(a) Organizing the Column. To facilitate control and scheduling, units will be organized into serials and march units and given an order of march. In determining the order of march, the march planner must consider the enemy situation and the desirable order of arrival of the units at the destination. Where dispersion is required, a unit may be organized into two or more columns, each assigned a different route.

(b) Using Reconnaissance Information. Results of the route reconnaissance will be used to select the route(s); determine the start point, critical points along the route(s), and the release point; and select the rate of march.

(c) Determining March Computations. March unit pass time is based on the strength, formation, and rate of march. The pass time of the marching columns, plus necessary time distance computations, will be used to determine the completion time of the march.

(d) Drafting of Road Movement Table. Using the completed march computations, a draft road movement tabic is compiled.

(e) Checking the Plan. Using the draft road movement table and a road movement graph, the movement plan is checked to ensure that it conforms to the directive of the higher headquarters and the battalion commander's instructions.

(5) Preparation and Issuance of the Road Movement Order. After the plan has been checked and approved by the commander, an operation order is prepared and issued. The operation order may be in written form or issued orally, and is accompanied by a road movement table, overlay and/or strip map, and appropriate administrative details.

(6) Road Movement Table. A road movement table, prepared as an annex to the operation order, provides serial commanders with arrival and clearance times at critical points along the route of march. It also provides the column commander with information as to the proposed location of elements of the column at various times.

(7) Overlay/Strip Map. An overlay serves the normal purpose and should show, as a minimum, the present location of units, route of march, critical points, and the new location of units at the destination. A strip map is a schematic diagram of the route of march and shows landmarks and critical points with the distances between them. A strip map may be issued as an annex to the road movement order, in addition to or in lieu of an overlay.

c. March Computations. Prior to issuance of the operation order, the S-3 must verify time and space computations as they provide him with the necessary data for the preparation of a road movement table.

(1) Time-Distance. Time distance (TD) is determined by dividing the distance to be traveled (D) by the rate of inarch (R):

TD = D/RWhere TD = hours D = distance in kilometers R = kilometers per hour

(2) Length of ColumnThe length of column (LC) is used to determine the pass time of a column. The sum total of the following two parts determines the length of column. (See pars, (a) and (b)).The two parts are the space occupied by the Marines alone (including the distance between Marines) and the sum of the distances between the units of the column (column gap).

(a) The length of column of Marines alone is determined by multiplying the number of Marines by the appropriate factor selected from the table below. The length of column does not include distances between units. LC Marines = No. of Marines x factor.

(b) The total distance in meters between units is determined as follows:

Step One. Determine the number of gaps between serials (total serials minus one).

Step Two. Multiply the number of serial gaps from Step One by the length (in meters) between respective units.

Step Three. Determine the number of gaps between march units (total march units minus one, minus the number of serial distances).

Step Four. Multiply the number of march unit gaps from Step Three by the length (in meters) between respective units.

Step Five. Add the totals from Step Two and Step Four to get the total meters for the column gap.

EXAMPLE:A battalion foot column is organized into 12 platoon-sized march units and 3 company-sized serials. REQUIRED: total column gap distances when there are 100 meters between serials and 50 meters between march units.

(3) Pass-Time.Pass-time (PT) is the time a unit takes to pass a specified point. For foot columns, the pass-time is determined by applying the following formula:PT (minutes) = LC x FACTOR (for appropriate rate of march).

PASS-TIME FACTORS - FOOT TROOPS

.0150 for 4.0 km/h
.0187 for 3.2 km/h
.0250 for 2.4 km/h
.0375 for 1.6 km/h

EXAMPLE: Determine the pass-time of a unit whose length of column is 1,500meters and is marching at a rate of 4 km/h.PT (min) = 1,500 x .0150 (the factor for 4.0 km/h) = 22.5 mm.

(4) Completion Time. Completion time is the time of day that a march will be completed. Completion time is determined by using the following formula: Completion Time = SP time + TD + PT + Scheduled Halts (other than normal hourly halts).

EXAMPLE: A column's starting point (SP) time is 0700.The time-distance is6 hours and 40 minutes. Pass-time of the column is 30 minutes. A 35-minutelunch halt has been scheduled. What is the completion time of the march?Employing the 24-hour clock system, the formula can be applied as follows for simplified addition of the times:

HrMin SP time................0700
TD.......................0640
PT.......................0030
Lunch halt..............0035

Completion time.........13 105or 1445

The march will be completed at1445.

(5) Experience Tables. Based on previous movements made by a unit, data is accumulated to facilitate march planning. Such data includes approximate pass-times lor various elements of the battalion. The S-3 can utilize these data rather than computing them each time a march is scheduled. Such experience tables serve to reduce the time required to complete the computation phase of march planning. Matter appropriate to the unit standing operating procedure should be integrated therein.

d. Road Movement Graph

(1) A road movement graph is a time-distance diagram used in planning, preparing, or checking road movement tables, and for controlling marches. The graph provides a visual representation of a march plan so that conflicts and discrepancies may be prevented in the planning stage, before congestion occurs on the route. It is not usually issued as a part of the order. Road movement graphs may be applied to small units, to movements of a single column, or to a large organization scheduling separate elements, moving by various means, with different rates of march, over one or more routes.

(2) To construct a road movement graph, use the following steps: (See fig. 3-2.)

Figure 3-2 is available in the illustration section at the end of the book

Step One. Determine the number of hours available for the march. Designate the lower left corner of a sheet of graph paper as the SP time or the earliest even hour before the march is to begin. Select a scale and plot the hours available in sequence from the left on the horizontal axis.

Step Two. Determine the distance to be moved in kilometers. Indicate the SP at the lower left corner of the graph sheet and, using an appropriate scale, plot the number of kilometers on the vertical scale from the SP to the release point. Indicate the location of critical points on the vertical scale.

Step Three. At the proper distance from the SP, draw a horizontal line indicating the location of the RP. Indicate by a vertical line the hour when the movement must be completed. Plot lines representing route restriction, if any, at the proper distances and times on the graph.

Step Four. Determine the pass-times of foot and motor elements of the column.

Step Five. Starting at the SP at the specified hour, plot the movement of the head of the leading element. If the rate of march is 3 kilometers per hour, the lead element will have moved 1.5 kilometers in 30 minutes, 3 kilometers in 1 hour, etc. Plot the trace of the lead man to the RP. The last Marine will cross the SP in pass-times after the lead man. Measure this time on the graph and plot the trace of the last Marine of the column. The lines describing the head and tail of a march serial are parallel. Indicate the time subsequent serials reach the SP and plot the trace of the head and tail. Check to see that the plan complies with all restrictions and orders. If there are any violations or conflicts, the plan may be made to conform by changing the order of march, the starting time of the serials, the planned rate of march, or the organization of the column.

e. Conversion to the Metric System. The adoption of the metric system for expressing linear distances requires a simple means of conversion.

(1) To accomplish this, use the following tables:

o Multiply inches x 2.54 to obtain centimeters. Multiply centimeters x .39 to obtain inches,

o Multiply yards x .91 to obtain meters. Multiply meters x 1.1 to obtain yards,

o Multiply miles x 1.6 (or 8/5) to obtain kilometers. Multiply kilometers x .62 (or 5/8) to obtain miles.

(2) Linear distances can be accurately measured on maps which have been converted to the metric system by using a straightedge graduated in centimeters (cm).The centimeter scale can be used in lieu of the graphic scale when the scale of the map is shown, as follows:

Map Scale:	One cm Equals:
1:500,000	5,000 meters
1:250,000	2,500 meters
1:100,000	1,000 meters
1:50,000	500 meters
1:25,000	250 meters

3104. INDIVIDUAL LOAD

Since the backbone of Marine ground combat forces is the infantry, tactical mobility is largely dependent on the marching ability of the individual Marine. The load that each Marine must carry is the greatest impediment to mobility. In general, the commander must make every effort to reduce the individual load of his Marines to the absolute minimum. While the load has the greatest effect on Marines marching on foot, the value of carrying the minimum essential supplies and equipment applies for all Marine forces.

a. Fighting Load. The fighting load consists of items of clothing, equipment, weapons, and ammunition that are carried by, and essential to, the effectiveness of the combat Marine and the accomplishment of the immediate mission. The fighting load should not weigh more than 40 pounds. The commander must ensure that his Marines' loads- -and his own- -be stripped to the bare minimum. The addition of small, light-weight items in excess of the limit should not be tolerated, since cumulatively, these items will take a toll in energy. Every extra pound a Marine must carry decreases his combat effectiveness.

b. Existence Load. The existence load consists of items other than those in the fighting load that are required to sustain or protect the combat Marine, that may be necessary for increased personal and environmental protection, and that are not normally carried by the individual. The existence load is normally carried by the Marine's unit. Marching, but not engaged in combat, a Marine may be expected to carry a maximum of 50 pounds of supplies and equipment.

c. Considerations

(1) The primary consideration is not how much a Marine can carry, but how much he can carry without impaired combat effectiveness-moral or physical. The combat strength of a unit is not counted simply in numbers of Marines, but in the numbers of willing and physically able Marines.

(2) It is generally better to risk temporary inconvenience from lack of health and comfort items than to exhaust Marines due to overloading.

(3) It is a fundamental truth that men become physically exhausted more quickly when under the stress of combat. Marines must be required to carry less into battle than they are conditioned to carry in training. Marines should be conditioned for carrying weight, but should be equipped in combat for fleetness of foot.

(4) A common mistake is to base the fighting load on the gear and supplies necessary to meet every contingency. The commander cannot reasonably expect the Marines to carry enough gear for every possible eventuality. The items to be carried must be based on reasonable expectations.
(5) It is the commander's responsibility to procure the transport to carry any additional gear. As a rule of thumb, a rifle company, or unit of similar size, requires one 5-ton truck and trailer in support to carry additional gear. In cold weather, or other conditions in which the necessary personal gear increases, this requirement will increase.
(6) The commander must ensure that the supply system provides, on a dependable and timely basis, the balance of essential supplies and equipment not carried by the unit. Marines must feel confident that they will be supported with the necessary supplies and equipment.
(7) In training, commanders must implant in their Marines the pride in operating under austere conditions. To effectively demonstrate the need for this spartan mentality, the commander must personally set the standard. Marines should be trained in field craft, foraging techniques, and the use of caches and field expedients. Maximum use should be made of captured stores.

3105. MARCH TECHNIQUES AND PROCEDURES

a. Organization. A command executing a march is organized into march units, march serials, and march columns, as necessary for control.
(1) March Units. A march unit is a unit of command which moves and halts at the command of a single commander. The march unit normally corresponds to one of the smaller Marine units such as a platoon or company.
(2) March Serials. A march serial consists of one or more march units organized under the senior officer and given a numerical or alphabetical designation to facilitate control. A serial is usually a battalion, but may be a company if the battalion is marching alone.
(3) March Columns. A march column is composed of elements of a command moving over the same route. It may be composed of one or more serials. To facilitate control, a column commander is designated. A column is normally a regiment or larger unit, but may be a battalion if marching alone.

b. Order of March. The order of march is determined by the mission, terrain, and the commander's desired order of commitment of units into action.

c. Control and Coordination Measures. The commander establishes initial control of the march by designating control measures in his march order. The most frequently used measures include:

o Start point and release point.
o Time at which head or tail of column passes the start point.
o Rate of march.
o Order of march.
o Route of march.
o Location of command post.
o March communications.

d. The Pacesetter. The pace-setter is an experienced individual carrying the same load as the majority of the Marines and marching from 4 to 10 meters at the head of the column. (See fig. 3-3.)The pace-setter's primary duty is to maintain the rate of march ordered by the column commander.

(l) Setting the Pace. The pacesetter does this by establishing the pace (length of step) and cadence (steps per minute) to obtain the prescribed rate of march. He should be of medium height so normal strides will be taken. Overstriding or understriding tends to tire leg muscles quickly and affects the combat efficiency of marching Marines. The officer marching at the head of the column supervises the pacesetter to ensure that he takes normal strides and maintains a uniform cadence.

Figure 3-3 is available in the illustration section at the end of the book

(2) Reducing Accordion Effect. Accordion effect occurs when the column alternately lengthens and shortens, causing the rear elements of the column to double time to maintain distance. It is caused by changes in the rate of march by the lead elements, usually after passing a slowing obstacle. To reduce this effect as much as possible, lead elements should slow their rate of march for a sufficient time after crossing obstacles to allow the rear elements to maintain the prescribed distance without running. The rear elements can signal the head when the last man has cleared the obstacle and the rate can be resumed. Since some accordion effect is nearly inevitable, the order of march should be periodically rotated to prevent units in the rear from becoming physically exhausted at the conclusion of the march.

e. Length, Rate and Pace

(1) **Length of March.** The length of march varies depending on terrain and weather, enemy situation, and physical and mental condition of Marines. The normal length of march for a 24-hour period is from 20 to 32 kilometers (12 to 20 miles) marching from 5 to 8 hours at a rate of 4 kilometers (2.5 miles) per hour (km/hmi/h).A march in excess of 32 kilometers in a 24-hour period is considered a forced march. Well-trained units that have been progressively trained will be able to achieve a unit goal of 60 miles in a 3-day period.

(2) Rate of March. The same factors of terrain, weather, load to be carried, and condition of Marines affect the rate of march. The commander considers these and selects a rate which will place his unit at the destination in the shortest possible time in condition to accomplish the mission. Unit standing operating procedure should state the standard rate of march for that unit over normal terrain for both road and cross-country movement and in both day and night. The commander then modifies this rate if the situation requires. Normally prescribed standards are as follows:

Figure is available in the illustration section at the end of the book

(3) Pace and Cadence. The normal pace is 30 inches. A pace of 30 inches and a cadence of 106 steps per minute result in a speed of 4.8 kilometers per hour or 3 miles per hour and a rate of 4 kilometers per hour or 2.5 miles per hour if a 10-minute rest halt per hour is taken. Since the pace of each man may vary, the cadence may have to be adjusted to maintain the prescribed rate. Slope of ground and surface footing affect pace. A normal cadence is relatively simple to maintain on level or gently rolling terrain on firm footing. Mud, sand, loose gravel, and hills all greatly affect pace.

f. Halts. Halts during a day's march are taken at regular intervals to rest personnel and adjust loads.Halts are specified by standing operating procedures or march order. Such factors as midday heat or enemy action may force the adoption of long midday halts or night marches. At long halts, each unit disperses to a previously selected location near the route of march.

(1) Normal Time and Frequency. Under normal conditions, a 15-minute halt is made after the first 45 minutes of marching. Following the first halt, a 10-minute halt is made after each 50 minutes of marching. Variations of this schedule are made when a scheduled halt time occurs when passing through a built-up area or when cover and concealment are required by the tactical situation and none is available. Observation posts may be established, if required for security of the unit during halts.

(2) Actions at Halts. All units in the column should be halted at the same time. At the halt signal, Marines should move to the side of the road, staying within the immediate vicinity of their unit. Marines should remove or loosen gear and sit or lie with feet elevated. Commanders inspect Marines and equipment, and corpsmen administer medical treatment, as required.

g. Communications. The four primary means of communications used in controlling foot marches are radio, visual, sound, and messenger.

(1) Radio. Radios are normally used for communications between platoons and higher headquarters in the march column. Radio transmissions should be held to the absolute minimum necessary for control and should be at minimum power required.

(2) Visual.. Visual communications may include flashlights, luminous markers, panel sets, pyrotechnics, and hand and arm signals. When contact with the enemy is not expected, the loss of light discipline from some of these signals may be a lesser security risk than the risk of disclosure from radio transmissions.

(3) Sound. Sound communications include voice and such signaling devices as horns and whistles. Some of their uses include assembling Marines at the conclusions of halts and providing warnings of air or chemical attacks.

(4) Messengers. Messengers are particularly effective during periods of radio silence and reduced visibility. Messages should be simple and brief to preclude mistakes by the messenger.

h. Security. Both active and passive measures are used against attack by enemy aircraft and indirect fire weapons during movement. Active measures include the use of air sentries and organic and attached weapons in accordance with the unit air defense standing operating procedure. Passive measures include the use of concealed routes and assembly areas, night marches, and extended interval between elements of the column. With imagination and planning, all of these measures can be practiced in training without detracting from the physical conditioning benefit of the march. In fact, such drills as rapidly taking cover during simulated attacks and then quickly resuming the march can prove physically challenging.

i. Reconnaissance. Reconnaissance determines in advance critical points along the march route such as bridges, fords, and obstacles in order that congestion or delay may be held to a minimum and local security provided. Paragraph 3103.C describes organization of the reconnaissance party.

3106.TACTICAL MARCHES

These marches are movements of Marines and equipment not in direct ground contact with the enemy but expecting early ground contact either en route or upon arrival at the objective. Some characteristics of this type of movement include security elements to the front, flanks, and, if necessary, to the rear of the march column. Additional considerations include the selection of concealed areas on defendable terrain for start and release points and rest stops. In training, infantry units of regimental size or smaller will normally derive the greatest benefit from tactical marches. The security requirements usually dictate a slower rate of march and, therefore, reduce the physical conditioning value for other types of units which would seldom move in combat with the expectation of early ground contact with the enemy. Two key points should be remembered regarding tactical marches in a training program.

a. Security. Although security requirements often slow the rate of march, it is a valid and important training objective for infantry units to work toward reducing march times while still practicing sound tactical security. March standing operating procedures, good land navigation skills, and simple but complete march orders contribute as much to this goal as physical conditioning. The idea that proceeding slowly enhances security is seldom true. Keeping the enemy off balance by rapid movement frequently is the best guarantee of security.

b. Training. In peacetime training, it is critical that support personnel who normally are located with infantry units in combat, participate completely with infantry units in forced march training. Not only must they not slow the infantry rate of march because of inferior conditioning, but they must be thoroughly familiar with the infantry unit's standing operating procedure for security on the march. Personnel from within the ground combat element such as artillery forward observer teams and combat engineer units, can be easily identified for this training and should be habitually associated with the same infantry units. Personnel from other elements, such as low-altitude air defense teams and radio battalion detachments, are more difficult to identify and earmark for support of the same infantry unit. Their normal mode of employment is in vehicles and their units frequently provide general support vice direct support. This does not, however, lessen their requirement to be prepared to provide support to dismounted infantry over terrain or in a tactical situation not conducive to vehicles. Such training needs to take place, and planning and cooperation among the elements of the MAGTF can overcome the obstacles.

3107.ADMINISTRATIVE MARCHES

These marches are movements of Marines and equipment made when no enemy interference is expected except from aircraft or indirect fire. During these marches, units are administratively grouped for ease of control and speed of movement. The term administrative foot marches should not be taken to imply that tactical considerations are disregarded. There are simply less stringent security measures used than when ground contact with the enemy is a distinct possibility. The normal formation is the route column with one file moving on either side of the road and with negligible security. Air sentries are used and dispersion between individuals and units is practiced. Although roads usually provide the fastest route, there may be reasons to conduct these marches cross-country (particularly as part of a training program). Administrative marches are the type of foot marches most practical for training programs for non-infantry units. These marches may also play a key role in the physical conditioning program of infantry units, because of the greater speeds possible due to lessened security requirements.

3108. FORCED MARCHES

A forced march is a march which covers a greater distance than 32 kilometer (20 miles) in a 24-hour period. Normally the extra required distance should be achieved by increasing the number of hours marched in a day beyond the norm (8 hours) rather than increasing the rate of march beyond the norm (4 kilometers per hour/2.5 miles per hour). However, there will be occasions when the situation or mission demands an increase in the rate of march.

a. Maximum Recommended Distances. The maximum recommended distances for forced marches are-56 kilometers (35 miles) in 24 hours; 96 kilometers (60 miles) in 48 hours; or 128 kilometers (80 miles) in 72 hours.

b. Sample Time Breakdown. A sample time breakdown for a forced march of 52 kilometers, beginning at first light is as follows:

Figure is available in the illustration section at the end of the book

3109. NIGHT MARCHES

Night marches are characterized by closed formations, more difficult control and reconnaissance, and a slower rate of march than day marches.

a. Control. Control is increased by reducing the distance between individuals and units, and by using connecting files to maintain contact between platoons and companies. Connecting files are normally constituted from personnel from the rear most march unit.

b. Safety on Roads. Because of reduced visibility, night marches on roads used by vehicles require attention to safety procedures to prevent accidents. Commanders should not think these safety measures apply only to peacetime training. In combat, nearly all vehicles operate with blackout lights at night, making visibility even more difficult than in garrison driving. If consistent with the tactical situation, the following measures can reduce the possibility of marchers being struck by vehicles:

(1) Use off-road trails and routes as much as possible.

(2) Guards to the front and rear of march units should be marked with strips of luminous tape and carry red-filtered flashlights. These measures are generally consistent with light discipline, while still affording reasonable warning to approaching drivers.

Section II. Conditioning Drills (One, Two, and Three)

3201. GENERAL

a. Description and Objective. Conditioning drills are calisthenic exercises. Each drill contains seven exercises organized and numbered in a set pattern. Each drill takes 15 minutes to complete. The objective of conditioning drills is to exercise all major muscle areas in order to develop strength, endurance, coordination, and flexibility.

b. Area and Equipment. Any level area is satisfactory for conduct of the drills. Drills One and Two contain ground exercises. If ground conditions are unsatisfactory, Drill Three should be used as it contains no ground positions. Usually, no equipment is required;however, if the group exceeds a platoon in size, an instructor's stand is necessary.

c. Formation. The extended rectangular formation is used in this drill. (See app. A.)

d. Starting Level and Progression. The starting level is six repetitions of each exercise. An increase of one repetition for each three periods of exercise in which the drill is performed is an acceptable rate of progression. This rate is continued until 12 repetitions can be completed. To maintain, continue the drill at 12 repetitions. To progress, move to 6 repetitions at a more difficult drill which exercises the same muscle group. Progression can also be gained by moving from Drill One to Drill Two, as Drill Two is more demanding.

e. Starting Positions. Starting positions vary with the exercise and are explained in each exercise.

f. Leadership. A principal instructor demonstrates and leads the drills. The instructor must be familiar with leadership techniques peculiar to conditioning drills to include the exercises, commands, counting cadence, cumulative count, formation, method of teaching the exercises, and utilization of assistant leaders.

g. Use With Other Programs. Conditioning Drills One, Two, and Three reach all major muscles of the body. They are easy to learn and to perform, and they are simple to administer and supervise. These features, coupled with the short time required for completion, the fact that no equipment is necessary, and adaptability to most areas of execution, make these drills possible in any programs.

3202. CONDITIONING DRILL ONE

This conditioning drill is similar to the calisthenics drill commonly known throughout the Marine Corps as the "Daily Seven."The "Daily Seven" may be used as a substitute for Conditioning Drill One.

a. Exercise l:High Jumper

(1) Starting Position. Feet separated shoulder width, knees flexed, body bent forward at the waist, arms aligned with the trunk and hips, elbows locked, palms facing, fingers extended and joined, head and eyes to the front.(See fig. 3-4, A.)(Elbows remain locked throughout the exercise.)

Figure 3-4 is available in the illustration section at the end of the book.

(2) Cadence. Moderate.

(3) Movement. A four-count exercise: at the count of—

(a) ONE -- Take a slight jump into the air, swinging the arms forward and up to shoulder level.

(b) TWO -- Take a slight jump into the air and swing the arms downward and back, returning to the starting position.

(c) THREE -- Take a vigorous leap into the air, swinging the arms forward and up to an overhead position, momentarily looking skyward, on returning to the ground the knees are flexed, head and eyes return to the front.

(d) FOUR--Repeat the action of count two.

b. Exercise 2: Bend and Reach

(1) Starting Position. Feet spread more than shoulder width, arms overhead, elbows locked, palms facing, fingers extended and joined, head and eyes to the front.(See fig. 3-4, B.)

(2) Cadence. Moderate.

(3) Movement. A four-count exercise: at the count of—

(a) ONE-Bend at the knees and waist, swing the arms straight downward and reach between the legs. Touch the ground as far to the rear as possible and look to the rear. (Elbows remain locked throughout the exercise).

(b) TWO-- Recover sharply to the starting position.

(c) THREE--Repeat the action of count ONE.

(d) FOUR-- Repeat the action of count TWO.

c. Exercise 3: Pushup

(1) Starting Position. Front leaning rest position. To assume this position there is a silent one-two count:on the silent count of one, assume the squatting position, heels together, elbows locked inside the knees, hands flat on the ground directly beneath the shoulders; on the silent count of two, thrust the legs to the rear, toes and heels together, body straight from head to heels.(Seefig. 3-4, C.)

(2) Cadence. Moderate.

(3) Movement. A four-count exercise: at the count of—

(a) ONE -- Flex the elbows lowering the body until the thick portion of chest touches the ground.

(b) TWO -- Raise the body until elbows are straight and locked.

(c) THREE-- Repeat the action of count ONE.

(d) FOUR -- Repeat the action of count TWO. (On returning to position of attention, the silent one-two count is used in reverse).

d. Exercise 4: Trunk Twister

(1) Starting Position. Feet are spread more than shoulder width apart, fingers placed behind neck, thumbs pointing downward, elbows back. (See fig. 3-4, D.)(Elbows remain well back throughout the exercise).

(2) Cadence. Slow.

(3) Movement. A four-count exercise: at the count of-

(a) ONE-- Keeping the knees locked and the back straight, bend forward at the waist sharply, with a slight recovery.

(b) TWO-- Twist the trunk to the left vigorously at the waist, keeping the elbow back. The left elbow is higher than the right.

(c) THREE-- Twist vigorously to the right, so the left elbow comes under the right.

(d) FOUR-- Straighten sharply to the starting position. NOTE: Do not attempt to touch the elbows to the knees on counts two and three.

e. Exercise 5: Squat Bender

(1) Starting Position. Feet are spread less than shoulder width apart, hands on hips, thumbs in small of back, elbows back. (See fig. 3-4, E.)

(2) Cadence. Moderate.

(3) Movement. A four-count exercise: at the count of—

(a) ONE-- Assume the squatting position, maintain balance on the balls of the feet, with trunk erect thrust arms forward to shoulder level, elbows locked, palms down.

(b) TWO-- Recover to starting position. Elbows are well back.

(c) THREE-- Keeping the knees locked, bend forward at the waist, touching the ground in front of the toes.

(d) FOUR-- Vigorously recover to the starting position.

f. Exercise 6: Body Twist

(1) Starting Position. On the back, arms extended sideward on the ground, palms down. The legs are raised to a near vertical position, feet together, knees locked.(See fig. 3-4, F.)

(2) Cadence. Slow-fast.

(3) Movement. A four-count exercise: at the count of-

(a) ONE-- Lower legs slowly to the left until they touch the ground near the left hand, keeping the knees straight and shoulders on the ground.

(b) TWO-- Recover the starting position by quickly raising the legs, keep knees straight and feet together.

(c) THREE-- Repeat movement of count ONE, except the movement is to the right side.

(d) FOUR-- Recover sharply to the starting position.

g. Exercise 7: Stationary Run

(1) Starting Position. Position of attention. (See fig, 3-4, G.)

(2) Cadence. Fast.

(3) Movement.

(a) At the command of execution, start running in place at double-time cadence, lifting the left foot first time cadence. Follow the instructor as he counts two repetitions of cadence;

e.g., 1, 2, 3, 4--1, 2, 3, 4. The instructor then gives informal commands such as FOLLOW ME, running on the toes and balls of the feet, keeping the back straight, speeding up the cadence to a sprint, raising the knees high, leaning forward at the waist, and pumping the arms vigorously.

(b) To halt the exercise, the instructor will count two repetitions of cadence as the left foot strikes the ground: 1, 2, 3, 4 - 1, 2, 3, HALT.

NOTE: When counting cadence, the instructor counts only as the left foot strikes the ground. The duration of the exercise is approximately 1 1/2 minutes.

3203. CONDITIONING DRILL TWO

a. Exercise 1: Jumping Jack

(1) Starting Position. Feet separated more than shoulder width, arms overhead. (See fig. 3-5, A.)

(2) Cadence. Moderate.

(3) Movement. A four-count exercise: at the count of~

(a) ONE-- Jump to position with the feet together and assume the squatting position, swinging the arms sideward and downward, placing the hands palms down on the ground, elbows locked inside the knees.

(b) TWO-- Recover to the starting position by jumping to the side straddle and swinging the arms sideward overhead.

(c) THREE-- Repeat the action of count ONE.

(d) FOUR-- Recover to the starting position.

b. Exercise 2: Turn and Bend

(1) Starting Position. Side straddle, arms overhead. (See fig. 3-5, B.)

(2) Cadence. Moderate.

(3) Movement. A four-count exercise: at the count of-

(a) ONE-- Turn the trunk to the left and bend forward over the left thigh, attempting to touch the fingertips to the ground outside the left foot. Keep the left knee straight. On successive repetitions attempt to touch farther and farther to the side.

(b) TWO-- Recover to the starting position.

(c) THREE-- Turn the trunk to the right and bend forward over the right thigh, trying to touch the hands to the ground outside the right foot. Keep the right knee straight.

Figure 3-5 is available in the illustration section at the end of the book.

(d) FOUR-- Recover to the starting position.

c. Exercise 3: Eight-Count Pushup

(1) Starting Position. Position of attention. (See fig. 3-5, C.)

(2) Cadence. Moderate.

(3) Movement. An eight-count exercise: at the count of—

(a) ONE-- Assume the squatting position, palms on the ground directly beneath the shoulders, elbows locked inside the knees.

(b) TWO-- Thrust the legs to the rear assuming the front leaning rest position.

(c) **THREE**-- Flex the elbows until the thick portion of the chest touches the ground.

(d) **FOUR**-- Raise the body on a straight plane until the elbows are locked.

(e) **FIVE**-- Repeat the action of count THREE.

(f) **SIX**-- Repeat the action of count FOUR.

(g) **SEVEN**-- Recover to the squatting position as in count ONE (elbows locked inside the knees).

(h) **EIGHT**—Return sharply to the position of attention.

d. Exercise 4: Turn and Bounce

(1) Starting Position. Feet spread more than shoulder width apart, arms sideward at shoulder level, palms up. (See fig. 3-5, D.)

(2) Cadence. Slow.

(3) Movement. An eight-count exercise: at the count of—

(a) **ONE**-- Turn sharply to the left as far as possible, then recover slightly.

(b) **TWO**-- Again turn to the left as far as possible and recover as in ONE.

(c) **THREE**-- Repeat the action of count TWO.

(d) **FOUR**-- Recover sharply to the starting position.

(e) **FIVE**—Turn sharply to the right as far as possible, then recover slightly.

(f) **SIX**—Again turn to the right as far as possible and recover as in FIVE.

(g) **SEVEN**-Repeat the action of count SIX.

(h) **EIGHT**-Return to the starting position.

NOTE: The head and hips remain to the front throughout the exercise and the knees and elbows are locked at all times.

e. Exercise 5: Squat Stretch

(1) Starting Position. Attention. (See fig. 3-5, E.)

(2) Cadence. Moderate.

(3) Movement. A four-count exercise: at the count of—

(a) **ONE**-- Squat, placing the hands on the ground about 12 inches in front of the feet.

(b) **TWO**— Keeping the fingertips on the ground, straighten the knees completely and raise the hips.

(c) **THREE**-- Recover to position ONE.

(d) **FOUR**-- Recover to the starting position.

f. Exercise 6: Leg Circular

(1) Starting Position. On the back, arms stretched sideward, palms down, feet raised foot from ground, knees straight. (See fig 3-5,F.)

(2) Cadence. Slow.

(3) Movement A four-count exercise: at the count of—

(a) **ONE--** Swing the legs as far as possible to the left, keeping the knees straight and the legs together.

(b) **TWO--** Swing the extended legs over head with the thighs as close as possible to the trunk.

(c) **THREE--** Swing the legs as far as possible to the right.

(d) **FOUR--** Recover to the starting position.

g. Exercise 7: Stationary Run

(1) Starting Position. Position of attention. (See fig. 3-5, G.)

(2) Cadence. Fast.

(3) Movement

(a) At the command of execution, start running in place at double-time cadence, lifting the left foot first. Follow the instructor as he counts two repetitions of cadence; for example: 1, 2, 3, 4--1, 2, 3, 4. The instructor then gives informal commands such as FOLLOW ME. Running on the toes and balls of the feet, keeping the back straight, speeding up the cadence to a sprint, raising the knees high, leaning forward at the waist, and pumping the arm vigorously.

(b) To halt the exercise, the instructor will count two repetitions of cadence as the foot strikes the ground: 1, 2, 3, 4-1,2, 3, HALT.

NOTE: When counting cadence, the instructor counts only as the left foot strikes the ground. The duration of the exercise is approximately 1 1/2 minutes.

3204. CONDITIONING DRILL THREE

a. Exercise 1: Side Straddle Hop

(1) Starting Position. Position of attention. (See fig. 3-6, A.)

(2) Cadence. Moderate.

(3) Movement. A four-count exercise: at count of-

(a) ONE-- Take a slight jump into the air, moving the legs sideward (more than shoulder width apart); at the same time, swing the arms overhead (to an overhead position) clapping the palms together.

Figure 3-6 is available in the illustration section at the end of the book.

(b) TWO--Take a slight jump into the air, swing the arms sideward and downward returning to the starting position.

(c) THREE-- Repeat the action of count ONE.

(d) FOUR-- Repeat the action of count TWO.

b. Exercise 2: Back Bender

(1) Starting Position. Standing, feet 12 inches apart, fingers placed behind the head. (See fig. 3-6, B.)

(2)Cadence. Slow.

(3) Movement. A four-count exercise: at the count of-

(a) ONE--Bend the upper trunk backward, raising the chest high, pulling the elbows back, and looking upward. Keep the knees straight.

(b) TWO-- Recover to the starting position.

(c) THREE-- Repeat the action of count ONE.

(d) FOUR-- Recover to the starting position.

c. Exercise 3: Squat Thrust

(1) Starting Position. Position of attention. (See fig. 3-6, C.)

(2) Cadence Moderate.

(3) Movement. A four-count exercise: at the count of—

(a) ONE-- Assume the squatting position; heels together, placing the hands flat on the ground, shoulder width apart, elbows locked and inside the knees.

(b) TWO-- Thrust the legs to the rear, assuming the front leaning rest position, body in line from head to toe, heels and toes together.

(c) THREE-- Return to the squatting position as in ONE.

(d) FOUR-- Return to position of attention.

d. Exercise 4: Side Bender

(1) Starting Position. Feet are spread more than shoulder width apart, arms are raised sideward and overhead, thumbs interlocked palms to front, fingers extended and joined, elbows locked. (See fig. 3-6, D.)

(2) Cadence. Slow.

(3) Movement. An eight-count exercise: at the count of—

(a) ONE-- Bend to left as far as possible, then recover slightly.

(b) TWO-- Again bend to the left as far possible, then recover slightly.

(c) THREE-Repeat the action of count TWO.

(d) FOUR-- Recover sharply to the starting position.

(e) FIVE-- Bend to the right as far as possible, then recover slightly.

(f) SIX-- Again bend to the right as far as possible, then recover slightly.

(g) SEVEN-- Repeat the action of count SIX.

(h) EIGHT-- Recover sharply to the starting position.

NOTE:Keep the elbows and knees locked throughout the exercise. The bend should occur to the side and not the front.

e. Exercise 5: Knee Bender

(1) Starting Position. Feet are spread less than shoulder-width apart, hands on hips, thumbs in small of back, elbows back. (See fig 3-6, E.)

(2) Cadence. Moderate.

(3) Movement. A four-count exercise: on the count of—

(a) ONE-Do a knee bend, lean trunk forward at the waist, thrust arms between legs until the extended fingers touch the ground palms to the ground, hands 6 inches apart.

(b) TWO~Recover sharply to the starting position.

(c) THREE-Repeat the action of count ONE.

(d) FOUR-Repeat the action of count TWO.

f. Exercise 6: Bottoms Up

(1) Starting Position. Front leaning rest position. A silent one-two count is used as in the pushups. (See fig. 3-6, F.)

(2) Cadence. Moderate.

(3) Movement A four-count exercise: at the count of-

(a) **ONE**-- With the weight on the hands, and knees locked, jump forward bringing the feet as close to the hands as possible; look to the rear.

(b) **TWO**--Keeping the knees locked, thrust the legs backward assuming the front leaning rest position.

(c) **THREE**-- Repeat the action of count ONE.

(d) **FOUR**-- Repeat the action of count TWO.

g. Exercise 7: Stationary Run

(1) Starting Position. Position of attention. (See fig. 3-6, G.)

(2) Cadence. Fast.

(3) Movement

(a) At the command of execution, start running in place at double time, lifting the left foot first. Follow the instructor as he counts two repetitions of cadence; for example: 1, 2, 3, 4--1, 2,3, 4. The instructor then gives informal commands such as FOLLOW ME, running on the toes and balls of the feet, keeping the back straight, speeding up the cadence to a sprint, raising the knees high, leaning forward at the waist, and pumping the arms vigorously.

(b) To halt the exercise, the instructor will count two repetitions of cadence as the left foot strikes the ground: 1, 2, 3, 4 - 1, 2, 3, HALT.

NOTE: When counting cadence, the instructor counts only as the left foot strikes the ground. The duration of the exercise is approximately 1 1/2 minutes.

Section III. Rifle and Log Drills

3301. RIFLE DRILL

a. Description and Objective. Rifle exercises are conditioning exercises performed with a rifle. Each drill contains six exercises (fig. 3-7) and they are numbered in a set pattern. The drill takes 15 minutes to complete. The objective of rifle drills is to exercise the arms, shoulders, and back muscles in order to develop strength and endurance, particularly in the upper body. In units without rifles, log drills may be substituted.

b. Area and Equipment. Any level area is satisfactory for conducting this drill. Each Marine completes these exercises from a standing position and no ground contact is required. Each Marine will need a rifle and, if the group exceeds a platoon in size, then the instructor will need an instructor's stand.

c. Formation. The extended rectangular formation is used in this drill. (See app. A.)
d. Starting Positions. Starting positions vary with the exercises and are explained in each exercise. As in all set conditioning drills, the command used to start the exercise is STARTING POSITION, MOVE. The following directions apply to rifle drill.
(1) In those exercises which start from the rifle downward position, on the command MOVE, execute port arms as prescribed in NAVMC 2691, Drill and Ceremonies Manual, and then assume the starting position. The command to return to the men to the position of attention at the conclusion of the exercise is POSITION OF ATTENTION, MOVE.
(2) In exercises which terminate in the rifle downward position, on the command MOVE, execute the position followed by order arms as prescribed in NAVMC 2691.
(3) In the exercises which terminate in a position other than the rifle downward position, Marines first assume the rifle downward position before executing port arms and order arms.
(4) These movements are executed without command. This procedure promotes uniformity, but precision is not expected. To be effective, rifle exercises must be strenuous enough to tire the arms, but not to the point where the arms cannot move with precision.

e. Leadership. A principal instructor demonstrates and leads the drill. He must be familiar with leadership techniques for conditioning exercises and the peculiar techniques for rifle drill.

3302. EXERCISES PERFORMED WITH RIFLES

The exercises of rifle drill are outlined in the following paragraphs,
a. Exercise 1: Foreup, Behind Back

(1) Starting Position. Rifle downward, feet together.(See fig. 3-7, A.)

Figure 3-7 is available in the illustration section at the end of the book.

(3) Movement. A four-count exercise: at the count of—
    (a) ONE-- Swing the arms forward and upward to the overhead position. Inhale.

(b) **TWO**-- Lower the rifle to the back of the shoulders. Exhale.

(c) **THREE**-- Recover to position ONE and inhale.

(d) **FOUR**-- Recover to the starting position and exhale.

b. Exercise 2: Lunge Side, Turn and Bend

(1) Starting Position. Rifle downward, feet together.(See fig. 3-7, B.)

(2) Cadence. Moderate.

(3) Movement. An eight-count exercise:at the count of—

(a) **ONE**-- Lunge sidewards to the left, swing the rifle forward and upward to the overhead position.

(b) **TWO**-- Turn the trunk to the left and bend forward over the left hip. At the same time, swing the rifle to a **low** horizontal in front of the left ankle.

(c) **THREE**-- Recover to position ONE.

(d) **FOUR**-- Recover to the starting position.

(e) **FIVE, SIX, SEVEN, and EIGHT**-- Repeat on the right side.

c. Exercise 3: Foreup, Back Bend

(1) Starting Position. Rifle downward, feet together.(See fig. 3-7, C.)

(2) Cadence. Moderate.

(3) Movement. A four-count exercise:at the count of—

(a) **ONE--** Swing the arms forward and upward to the overhead position.

(b) **TWO**-- Bend backward, emphasizing the bend in the upper back. The face is up.Keep the knees straight.

(c) **THREE**-- Recover to position ONE.

(d) **FOUR**-- Recover to the starting position.

d. Exercise 4:Up and Forward

(1) Starting PositionRifle downward, feet together.(See fig. 3-7, D.)

(2) Cadence. Fast.

(3) Movement. A four-count exercise:at the count of—

(a) **ONE**-Swing the arms forward and upward to the overhead position.

(b) **TWO-** Swing the arms forward to shoulder level.

(c) **THREE**-Recover to position ONE.

(d) **FOUR**-Recover to the starting position.

e. Exercise 5: Foreup, Full Squat

(1) Starting Position. Rifle downward, feet in narrow stance. (See **fig.** 3-7, E.)

(2) Cadence. Moderate.

(3) Movement. A four-count exercise: at the count **of**—

(a) **ONE**-- Swing the arms forward and upward to the overhead position.

(b) **TWO**-- Swing the arms down to shoulder level and assume the squatting position.

(c) **THREE**-- Recover to position ONE.

(d) **FOUR**-- Recover to the starting position.

f. Exercise 6: Arms Forward, Side Bend

(1) Starting Position. Sidestraddle, regular stance, rifle forward.(See fig. 3-7, F.)
(2) Cadence. Moderate.
(3) Movement. A four-count exercise:at the count of-

(a) **ONE**-- Bend the trunk to the left. Keep the knees straight.
(b) **TWO**-- Recover to the starting position.
(c) **THREE**-- Bend the trunk to the right. Keep the knees straight.
(d) **FOUR**-- Recover to the starting position.

NOTE:Keep the rifle on the same level as the shoulders throughout the exercise.

3303. LOG DRILL

a. Description and Objective. Log exercises are conditioning exercises performed with a log. Each drill contains six exercises and they are numbered in a set pattern. The drill takes 15 minutes to complete. The objective of log drills is to develop strength and muscular endurance and, in this instance, under maximum loads. Log exercises also develop teamwork. Log may be used in lieu of conditioning drills after the Marines have become somewhat conditioned.

b. Area and Equipment. Any level area is satisfactory for conducting this drill. Each Marine completes these exercises from a standing position and no ground contact is required. If the group exceeds a platoon in size, then the instructor will need instructor's stand. Each six-person group or eight-person group will need a log. The logs should be from 6 to 8 inches in diameter. They may vary in length from 14 feet (for 6 people) to 18 feet (for 8 people).The logs should be skinned, smoother, and dried. The 14-foot logs should weigh approximately 300 pounds and the 18-foot logs, approximately 400 pounds. Rings should be painted on the logs to indicate each person's position. When not in use, the logs should be stored on a rack to keep them off the ground.

c. Formation. All the Marines assigned to the same log team should be about the same height at the shoulders. The recommended method of dividing the platoon is to have the Marines form a single file or column with short people to the front and tall people to the rear. Have the Marine assume their positions in the column according to shoulder height, not head height. When they are in position, they are given the command COUNT OFF BY SIXES (OR EIGHTS), COUNT OFF, to divide them into six- or eight-person log teams. Each team in turn, can then proceed to the log rack, shoulder a log, and carry it to the designated exercise area. The log teams form in columns in front of the instructor. With the Marine holding the log in the chest position, have them face the instructor and ground the log at least 10 yards between columns and 10 yards between log teams within the columns.

d. Starting Positions. The Marines fall in, facing the log, with their toes about 4 inches from it. The basic starting positions and commands are as follows:(See fig. 3-8.)
(1) **RIGHT-HAND STARTING POSITION, MOVE.** It the command MOVE, move the left foot 12 inches to the left, and lower the body into a flat foot squat. Keep the back straight, head up, and arms between the legs. Encircle the far side of the log with the left hand. Place the right hand underneath the log.(See fig. 3-8, A.)

(2) LEFT-HANDED STARTING POSITION, MOVE. These commands are executed in the same manner as in paragraph a. except that the left hand is underneath the log and the right hand encircles its far side. (See fig. 3-8, B.)

Figure 3-8 is available in the illustration section at the end of the book.

(3) RIGHT SHOULDER POSITION, MOVE. At the command MOVE, pull the log upward in one continuous motion to the right shoulder. At the same time, move the left foot to the rear and stand up, facing left. Balance the log on the right shoulder with both hands.(See fig. 3-8, C.)This movement cannot be performed from the left-hand starting position because of the position of the hands.

(4) LEFT SHOULDER POSITION, MOVE. These commands should be given from the left hand starting position. At the command MOVE, pull the log upward in one continuous motion, to the left shoulder. At the same time, move the right foot to the rear and stand facing right. Balance the log on the left shoulder with both hands.(See fig. 3-8, D.)This movement cannot be performed from the right-hand starting position.

(5) WAIST POSITION, MOVE. From the right hand position pull the log waist high. Keep the arms straight and fingers laced underneath the log. The body is inclined slightly to the rear and the chest is lifted and arched. (See fig. 3-8, E.)

(6) CHEST POSITION, MOVE. This command should be given after the waist position has been assumed .On the command MOVE, shift the log to a position high on the chest, bring the left arm under the log and hold the log in the bend of the arms.(See fig. 3-8, F.)Keep the upper arms parallel to the ground.

(7) To move the log from the right shoulder to the left shoulder, the command is: LEFT SHOULDER POSITION, MOVE. On the command MOVE, push the log overhead and lower it to the opposite shoulder.

(8) To return the log to the ground from any of the above positions, the command is: STARTING POSITION, MOVE. At the command MOVE, slowly lower the log to the ground. The hands and fingers must be kept from under the log.

e. Leadership. A principal instructor demonstrates and leads the drill. He must be familiar with the leadership techniques for conditioning exercises and the peculiar techniques for log drill.

3304. EXERCISE PERFORMED WITH LOGS

The exercises of log drill are outlined in the following paragraphs. Figure 3-9 graphically explains log drill.

a. Exercise l:Two-Arm Pushup

(1) Starting Position. Right or left shoulder position. Regular stance. (See fig. 3-9, A.)

(2) Cadence. Moderate.

(3) Movement. A four-count exercise:at the count of—

- **(a) ONE--** Push the log overhead until the elbows lock.
- **(b) TWO--** Lower the log to the opposite shoulder.
- **(c) THREE--** Repeat the action of count ONE.
- **(d) FOUR--** Recover to the starting position.

b. Exercise 2: Forward Bender

(1) Starting Position. Chest position. Regular stance.(See fig. 3-9, B.)

(2) Cadence. Moderate.

(3) Movement. A four-count exercise:at the count of-

(a) ONE-- Bend forward at the waist, keeping the back and legs straight.

(b) TWO-- Recover to the starting position.

(c) THREE-- Repeat the action of count ONE.

(d) FOUR-- Recover to the starting position.

c. Exercise 3: Straddle Jump

(1) Starting Position. Right or left shoulder position, feet together, fingers interlaced on top of the log.(See fig. 3-9, **C.)**

(2) Cadence. Moderate.

(3) Movement. A four-count exercise:at the count of—

(a) ONE-- Jump to a side-straddle. Pull down on the log with both hands to keep it from bouncing on the shoulder.

(b) TWO-- Recover to the starting position.

(c) THREE-- Repeat the action of count ONE.

(d) FOUR-- Recover to the starting position.

Figure 3-9 is available in the illustration section at the end of the book.

d. Exercise 4: Side Bender

(1) Starting Position. Right shoulder position, feet regular stance.(See fig. 3-9, D.)

(2) Cadence. Moderate.

(3) Movement. A four-count exercise:at the count of—

(a) ONE-- Bend sideward to the left as far as possible, bending the left knee.

(b) TWO-- Recover to the starting position.

(c) THREE-- Repeat the action of count ONE.

(d) FOUR-- Recover to the starting position.

(4) After completing the required number of repetitions, change shoulders and execute an equal number of repetitions to the other side.

e. Exercise 5: Knee Bend

(1) Starting Position. Right or left shoulder position. Narrow stance. Fingers interlocked on top of the log.(See fig. 3-9, E.)

(2) Cadence. Slow.

(3) Movement. A four-count exercise:at the count of—

(a) ONE-- Flex the knees to a quarter-squat position.

(b) TWO-- Flex the knees to a half-squat position.

(c) THREE-- Lower the body to a three-quarter squat position.(Lean slightly forward.)

(d) FOUR-- Recover to the starting position.

NOTE:Pull forward and downward on the log throughout the exercise.

f. Exercise 6: Overhead Toss

(1) Starting Position. Right or left shoulder position, regular stance. The knees are bent to a quarter-squat.(See fig. 3-9, F.)

(2) Cadence. Moderate.

(3) Movement. A four-count exercisc:at the count of-

(a) ONE-Straighten the knees and toss the log into the air approximately 12 inches overhead. Catch the log with both hands and lower it toward the opposite shoulder. As the log is caught, lower the body into a quarter-squat.

(b) TWO-- Again toss the log into the air and when caught return it to the original shoulder.

(c) THREE-- Repeat the action of count ONE.

Section IV. Grass Drills

3401. GENERAL

a. Description and Objective. Grass drills are extremely strenuous exercises and are performed at top speed for only short periods of time. No cadence is counted but the Marines continue to execute the multiple repetitions of the command until the next command is given. The grass drill consists of two drills:Drill One and Drill Two. Each drill contains six exercises. The objective of the drills is to decrease reaction time, to develop cardiovascular endurance, and to provide a vigorous workout for all major muscles.

b. Area and Equipment. Any level area suitable for ground contact and of a size to accommodate the group is adequate. No equipment is needed.

c. Formation. All movements are executed in place. The extended, rectangular formation is recommended for a platoon- or company-size unit. The circle formation is suitable for groups of squad or section size. At the beginning of an exercise program, 2 to 3 minutes of grass drills will insure a good workout.

d. Progression. Progression is gained by gradually increasing the length of time devoted to the drills. As the physical condition of the Marines improves, the periods should be gradually lengthened to 5 minutes. As the second drill is more difficult than the first, some progression can be attained by initially executing Drill One; then as the program and the Marines progress, introduce Drill Two. To extend the duration of the drill, it may be necessary to repeat the drill.

e. Starting Position

(1) The drills are started from the GO position. Other basic positions are FRONT, BACK, and STOP. (See fig. 3-10, A.)

(a) GO. Running in place (top speed): on the toes and balls of feet, knees raised high, arms pumping, body bent forward at waist.

(b) FRONT. Prone position: elbows bent (along body), palms flat on ground directly under the shoulders, legs together and straight.

(c) BACK. Supine position (flat on back): arms extended near side on ground with palms down, legs together and straight, feet toward the stand or instructor.

(d) STOP. Football lineman stance: feet spread and staggered, left arm across left thigh, right arm straight, knuckles on ground, head up, back parallel with ground.

(2) To assume the FRONT or BACK position from the STANDING, GO, or STOP position, vigorously get into the prescribed position as quickly as possible. (See fig. 3-10, B.)

(3) To change from the FRONT to the BACK position, quickly do a pushup, move the feet several short steps to the right or left, lift the arm on the side toward which the feet move, and thrust the legs vigorously to the front. (See fig. 3-10, C.)

Figure 3-10 is available in the illustration section at the end of the book.

(4) To move from the BACK to the FRONT position, sit up quickly, place both hands on the ground to the right or the left of the legs. Move the feet several short steps to the rear on the side opposite the hands. When the feet are opposite the hands, thrust the legs vigorously to the rear and lower the body to the ground.(See fig. 3-10, D.)

f. Leadership. A warm up activity of lesser intensity should proceed grass drill. During the instructional phase and conduct of these drills, the following points should be applied.

(1) The instructor executes only GO and STOP with the Marines. This allows the instructor to supervise the drill.

(2) The commands for grass drills are given in rapid succession without the usual preparatory command.

(3) To prevent confusion, the instructor should give the commands sharply to distinguish them from comments or encouragement.

(4) As soon as the Marines know the drill, they should respond to the instructor's commands and perform all exercises vigorously and as rapidly as possible. All exercises are executed continuously until the next command is given. Insist on top speed performance; anything less is not effective.

(5) The commands peculiar to each exercise are identical to the name of the exercise.

(6) Marines are not to be required to assume the position of attention once the drills are started. To halt the drill for instructions or for rest, the command UP is used. At this command, the Marines assume a relaxed standing position. Do not demand formality. At the conclusion of a fast and vigorous 5-minute grassdrill, it is physically impossible for people to stand at attention.

(7) The sequence of commands for the execution of grass drills should occur as follows:Drill One. GO, FRONT, Bouncing Ball; GO, BACK, Bicycle; GO, Full Squatter; GO, BACK, Situps; GO, FRONT, Mountain Climber; GO, FRONT, Roll Left; GO, STOP, UP.

g. Use With Other Programs. Since grass drills can be executed in a short period of time, they may be executed where only a few minutes are available for exercise or in conjunction with another type of activity. Grass drills are an excellent substitute for running when time is a factor.

3402. GRASS DRILL ONE AND TWO

a.Grass Drill One

(1) Bouncing Ball. From the FRONT position, push up, supporting the body on the hands (shoulder-width apart) and feet. (See fig. 3-11.)Keep the back and legs in line and the knees straight. Bounce up and down by a series of short, upward springs from the hands, hips, and feet simultaneously,

(2) Bicycle. From the BACK position, raise the legs and hips. Keep the elbows on the ground and support the hips with the hands. Move the legs vigorously as if pedaling a bicycle.

(3) Full Squatter. From the STOP position, assume a full knee bend, the feet on line, hands on hips. Bounce up and down in place by short, bouncing jumps.

Figure 3-11 is available in the illustration section at the end of the book.

(4) Situps. From the BACK position and with arms stretched overhead, sit up, reach forward, and touch toes. Return to the supine position.

(5) Mountain Climber. From the STOP position, place both hands on the ground directly under the shoulders. Thrust the right leg to the rear, knee straight. The left foot should be close to the left hand, the left knee outside the left arm. Shift the weight to the hands, thrust off with the rear (right) foot and bring that foot up close to the right hand, the right knee outside the right arm. At the same time, thrust the left leg vigorously to the rear, knee straight. Continue at a fast cadence, alternating the legs.

(6) Roll Left. From the BACK or FRONT position, make one complete roll in the direction commanded. On completing the roll, return to the FRONT or BACK position.

b. Grass Drill Two

(1) Legs Over. .From the BACK position and with arms stretched overhead, palms up, raise the legs upward and then swing them backward over the head until the toes touch the ground behind the head. Return legs to the starting position. (See fig. 3-12.)

(2) V-Up and Touch Toes. .From the BACK position, raise the legs with the knees straight, sit up until the trunk and legs form a V, and touch the toes with the hands. Return to the BACK position.

(3) Rocker. .In the FRONT position, clasp the hands behind the back, arch the body, holding the head back. Start rocking, using the front part of the trunk as a rocker.

(4) Bounce and **Clap Hands**. The procedure is the same as for bouncing ball, but while in the air, clap the hands. This requires a more vigorous bounce or spring. (See fig. 3-12.)

(5) Leg Spreader. .From the BACK position, raise the legs so that the heels are 10 to 12 inches from the ground, spread them apart as far as possible, then close them together. Open and close legs as rapidly as possible.

(6) Forward Roll. .For forward roll from the STOP position, place both hands on the ground, tuck the head, and do one complete forward roll, keeping the legs tucked as you roll, and come back to the STOP position.

Figure 3-12 is available in the illustration section at the end of the book.

Section V.Guerrilla Exercise

3501.GENERAL

a. Description and Objective. Guerrilla exercises are individual exercises of various basic skills that are performed rapidly while moving forward in a circle formation. There are two tables of guerrilla exercises. Each table takes 15 minutes to complete. The objective of this exercise is to increase strength and endurance, aid flexibility, and develop coordination.

b. Area and Equipment. Any level area is suitable for conducting guerrilla exercises. No ground contact is required other than the hands. There is no equipment requirement.

c. Formation

(1) The circle formation (app. A) is used for guerrilla exercises. Each platoon forms its own circle and engages in guerrilla exercises under a platoon instructor. If the platoon exceeds 30 Marines, double or concentric circles may be used.

(2) When the circle is formed, the instructor steps into the center of the circle and moves clockwise in a small circle. He commands:QUICK TIME, MARCH, 1-2-3-4. (Rapid cadence of approximately 130 counts per minute. Cadence and step are maintained between exercises.)

(3) To reform the platoon after completing guerrilla exercises, the instructor halts the Marines and places the base man or platoon guide where he wishes and commands:

(a) BASE MAN (or platoon guide),POST.

(b) FALL OUT AND FALL IN ON THE BASE MAN (or platoon guide).

d. Progression. Progression may be attained by moving from table I to table II or by shortening the quick time marching periods between exercises and performing all exercises a second time.

e. Leadership

(1) To execute the exercises, the Marines continue at quick time while the instructor simultaneously explains and demonstrates the exercise to be performed, and then commands the Marines accordingly. In each instance, the preparatory command will be the name of the exercise and, in all instances, the command of execution will ***MARCH***.. *To* terminate each exercise, the command *is* ***QUICK TIME, MARCH.*** *The* Marines immediately pick up the step as the instructor counts cadence.

(2) Unless specified differently, each exercise should be continued for 20 to 40 seconds depending upon the vigor of the exercise. The leader can determine the duration of each exercise by observing its effect upon the Marines.

(3) To form for double guerrillas, the commands for pairing the Marines (who are in circle formation) are-

(a) PLATOON, HALT.

(b) FROM (designate an individual), BY TWO'S, COUNT OFF(.Examplel-2; 1-2; 1-2; etc.)

(c) EVEN NUMBERS MOVE UP BEHIND ODD NUMBERS(At this time, adjust pairs according to height and weight.)

(d) YOU ARE NOW PAIRED UP FOR DOUBLE GUERRILLAS. To change the Marines position, merely command CHANGE.)
(e) FORWARD, MARCH.

f. Place in the Program. Many Marines have not had the opportunity to perform the simple skills involved in guerrilla exercises. The conduct of these exercises is a simple matter since they can be performed easily and quickly in almost any situation. The tables of exercise are applicable to all personnel. The tables can constitute a station within a 1-hour period or be completed within a separate 15-minute period.

3502.GUERRILLA TABLES

a. Table I
(1) Double Time.(See fig. 3-13, A.)Hold arms in the thrust position. Execute a double-time run, maintaining the circle formation and the prescribed distance between your fellow Marines. Duration-1 minute.
(2) All Fours.(See fig. 3-13, B.) Face downward. Support the body with the hands and feet. Walk forward hands first.
(3) Crab Walk.(See fig. 3-13, C.) Get in the sitting position, face upward, and lift the hips. Support the body with the hands and feet. Walk forward feet first.
(4) Squat Walk.(See fig. 3-13, D.) Assume a full knee bend position. Grasp the ankles (left ankle with the left hand, right ankle with the right hand). Walk forward.
(5)Broad Jump.(See fig. 3-13, E.) Jump forward on both feet in a series of broad jumps. Swing the arms vigorously to assist the jumps.
(6)Toe-Touch Walk.(See fig. 3-13. F.) Walk forward, bending at the waist and touching one hand to the toe of the opposite foot while it is on the ground. Raise the trunk to the vertical position between steps. Keep the knees straight.
(7)Bottoms-Up Walk.(See fig. 3-13, G.) Assume the front leaning rest position and move the feet toward the hands in short steps, keeping the knees locked. When the feet are as close to the hands as possible, walk forward on the hands to the front leaning rest position.
(8) Straddle Run.(See fig. 3-13, H.) Run forward, leaping to the right from the left foot and to the left from the right foot.
(9) Fireman's Carry .(See fig. 3-13,1.)
(IO)Single-Shoulder Carry(See fig. 3-13, J.) Two men execute the carries as indicated by the diagram. No. 1 man executes one type; No. 2 man executes the other.

Figure 3-13 is available in the illustration section at the end of the book.

b. Table II
(1) Double Time.(See fig. 3-14, A.)Hold arms at the thrust position. Execute a double-time run, maintaining the circle formation and the prescribed distance between your fellow Marine.Duration-1 minute.
(2) Toe-Grasp Walk.(See fig. 3-14, B.) Bend forward and grasp toes. With knees slightly bent, walk forward.

(3) Hand-Kick Walk..(See fig. 3-14, C.) Walk forward, kicking the moving foot upward on every step. At the same time, lean forward and touch the elevated toe with the hand of the opposite arm.

(4) Pike Juinping.(See fig. 3-14, D.) Jump forward and upward from both feet, keeping the knees straight, and at the same time, swing the legs forward and touch the toes with the hands at the top of each jump.

(5) Squat Jump.(See fig. 3-14, E.) Leap forward from the squatting position, with the hands on the ground and the arms between the legs. Land on the ground with hands and legs extended. Bring up the legs to the squatting position.

(6) Steam Engine..(See fig. 3-14, F.) Lace the fingers behind the neck and walk forward in the following manner: as the left leg moves forward, raise the knee high, bend the trunk forward, and touch the outside of the right elbow to the outside of the knee. Then lower the left leg and step forward on the left foot and raise the right leg. Repeat with the right leg and left elbow.

(7) Knee-Touch Walk.(Scc fig. 3-14, G.) Walk forward, bending the knees and touching the ground on each step. The knees are bent and straightened on each step.

(8)Hobble Hopping.(See fig. 3-14, H.) Hold foot behind back with opposite hand and hop forward. On the command CHARGE, grasp the opposite foot with opposite hand and hop **forward.**

(9) Cross Carry.(See fig. 3-14,1.)(10)Saddle Back Carry. (See fig. 3-14, J.) Two Marines execute the carries as indicated in the diagram. No. 1 Marine executes onetype;No. 2 Marine executes the other.

Figure 3-14 is available in the illustration section at the end of the book.

Section VI. Running and Orienteering

3601. RUNNING

a. General

(l)Description. The general form and technique for all types of running are fairly constant. (See fig. 3-15.)The head is erect, body slightly forward without bending at the waist, and the arms are at a loose thrust position alternating from front to rear in straight planes. A cross-body arm movement wastes energy .The movement of the legs and feet will be discussed in subsequent paragraphs dealing with the different types of running. Of primary importance is the fact that in all types of running, the toes should be pointed straight ahead. Toeing out is a common error in both running and walking and should be an item of individual correction.

(2) Objective. The objective of running is to develop cardiovascular endurance. Despite the fact that Marines have developed their muscle structure and the strength of their muscle tissue, unless they have developed cardiovascular endurance to a satisfactory degree, they are not entirely physically fit or combat ready. Cardiovascular endurance (wind) depends on the efficiency of the lungs and heart. The efficiency of the lungs and heart depend on the amount of oxygen the lungs can absorb with each breath inhaled and the amount of carbon dioxide the lungs can expel. The process of absorbing oxygen and expelling carbon dioxide (cardiopulmonary process) is performed by the blood that circulates through the lungs. The condition of this process will determine the amount of effort a person can exert over a period of time. Running is one of the best activities to develop this vitally important endurance.

b. Running Skills. In the development of running skills, individuals may require instruction to improve their proficiency. Some important skills to consider are-

Figure 3-15 is available in the illustration section at the end of the book.

(1) Action of the Arms. Arm action is important. Check to see that arms are held loosely and that the action is relaxed. The faster the run, the more rapid the arm action.

(2) Breathing. Allow the individual to breathe through the mouth as the body demands a large supply of oxygen. Oxygen can be inhaled in greater quantities through the mouth.

3602. TYPES OF RUNNING

a. Double Time

(1) Description and Objective. Double timing is marching at the rate of 180 steps per minute, each step being 36 inches in length. It takes practice to double time with precision in formation. The Marines should keep in step, placing their feet flat on the ground. This, however, should not be a stamping motion, but should be done with as slight a jolt as possible. Double timing is like a jog, the difference being that in a jog the feet are lifted well off the ground and the running motion is bouncy. In double timing, the feet skim the ground and there is no bounce to the ran. Double timing is a vehicle for teaching proper running form and for the development of the cardiovascular system.

(2) Area and Equipment. This type of running can be completed over a variety of surfaces. Usually a Marine uses a field or road. There is no equipment requirement.

(3)Progression. There is no set standard for alternating quick time and double time in the early conditioning of Marines. A general rule is to begin with enough quick-time marching to ensure a thorough warming up, then double time about 100 paces. Change again to quick time until the individuals have made a reasonable recovery from the running, then double time another 100 paces. The amount of double time can be increased and the quick time decreased from week to week, until the individuals are double timing about 1800 yards. This type of training should be given at least twice a week, but by no means is it adequate as the sole means of conditioning.

(4) Leadership

(a) The instructor should be to one side of the column or group and toward the rear so there is a full view of all Marines. Inexperienced instructors have a tendency to supervise from a position too far forward.

(b) Select an individual who can maintain the proper cadence to act as the guide during double-time running.

(c) There are several ways for the instructor and group to count cadence while double timing. If not contrary to local policy, learn several methods and use them for variety.

b. Wind Sprints

(1) Description and Objective. This type of running involves a series of 30- or 40-yard dashes, usually conducted in successive waves of squads. Each squad is in line and the squad leader on the right flank. Wind sprints assist indeveloping speed and cardiovascular endurance. Any flat and level area may be used which will permit the squad to form a line and run the required distance.

(2) Progression. One or two 30-yard sprints will be adequate at the beginning. As time passes, sprints can be lengthened and up to six or seven sprints may be used.

(3) Leadership

(a) At the command READY (given by squad leader), each runner assumes the sprinter's starting position. At the command GO, the squad sprints approximately 30 yards, takes 10 yards to stop, and lines up immediately with the squad leader who repeats, READY, GO, and again the squad sprints. At the conclusion of the third sprint, the squad waits until all the squads of the platoon have made three sprints. Then they all line up and the squad leaders conduct three more wind sprints in the opposite direction.

(b) Valuable time is gained by having each squad ready to go when the preceding squad has moved off its second sprint mark.

c. Cross-Country Running

(1) Description and Objective. Cross-country running is a distance run conducted on a course laid out along roads, across fields, over hills, through woods, and on any irregular ground. The cross-country ran may be utilized as a conditioner or as a competitive event; the objective is to cover the distance in the shortest possible time. The course should be 2 to 2.5 miles in length and be laid out to avoid heavy vehicular traffic. The course should be marked by directional arrows until the runners know the course. These runs build leg muscles, increase lung capacity, and develop endurance. Any local area of varied terrain is suitable.

(2) Progression. In the mass training of a large group, leaders should be stationed at the head and the rear of the column and should make every effort to keep the individuals together .After determining the abilities of the individuals in cross-country running, it is advisable to divide the unit into three groups. The poorest conditioned group starts first, and the best conditioned group, last. The starting time of the groups should be staggered so that all of them finish about the same time. In preliminary training, the running is similar to ordinary road work in that it begins with rather slow jogging, alternating with walking. The speed and distance of the run is gradually increased. As the condition of the individual improves, occasional sprints may be introduced. At first the distance run is from 1/2 to 1 mile. It is gradually increased to 2 or 2.5 miles. Well-conditioned personnel can run 2 to 2.5 miles within a 15-minute period.

(3) Leadership. Marines should not be required to take part in distance running until they have been through a progressively scheduled training program which requires a considerable amount of running. Cross-country runs should be scheduled occasionally to provide variety in the program. Cross-country running has the advantage of allowing mass participation. Interest can be stimulated by putting the runs on a competitive basis. As a single activity, short cross-country runs can be scheduled once a week, gradually increasing the distance as the physical conditioning improves.

(4)Use With Other Programs. Cross-country running can be combined with other activities such as conditioning exercises.

d. Fartlek Training

(l)Description and Function. Fartlek training is various running exercises conducted along an intense combat-like course. The running is conducted to overwork the lungs, allowing only partial recovery which is followed by another intense period of overload. This sequence is repeated for the duration of the workout. Fartlek training is a useful combat training method and general cardiovascular conditioner. The entire unit must be in uniformly good to excellent condition to qualify for this training.

(2) Progression. The following is an example of one fartlek session:

(a) Warm up by stretching 3-5 minutes.

(b) Warm up running easily 5-10 minutes.

(c) Run at a fast, steady pace for 3/4 to 1 3/4 miles (dependent on terrain).

(d) Walk/jog at a moderate pace for 5 minutes (recovery).

(e) Run easily sprinting 15 to 20 meters occasionally.

(f) Run full speed uphill for 175 to 200 meters.

(g) Warm down by running easily for 1/2 to 1 mile.

(h) Warm down by stretching 3 to 5 minutes.

(3)Leadership. To add variety, try incorporating combat movements, terrain association, a series of exercises or carrying table of equipment weapons (hit and roll). This will take some initiative in setting the course, but it allows for variety. In this manner, the fartlek course can be used as a training session or part of one. It works extremely well with highly motivated, competitive Marines.

3603.ORIENTEERING

Orienteering is land navigation over a prescribed course as a timed, competitive event. It is an excellent way to integrate land navigation training with physical conditioning and requires Marines to be skilled in both areas. Commanders adjust the training as needed to change the degree of difficulty or to emphasize either navigation or conditioning. Taken to the extreme, an orienteering exercise is conducted in full combat gear over rugged terrain against aggressors. Orienteering requires the following physical skills: endurance, speed, strength, and all-around conditioning.

Section VII. Circuit Training

3701. GENERAL

a. Description and Objective. A strength circuit consists of a series of stations where individuals in small groups exercise vigorously for a short period of time and then move (on signal) to the next station where a different form of exercise is conducted. This rotation continues until all groups move through all stations. Strength circuits contain no set or specific types of exercise stations within the circuit. The objective of circuit training is to develop strength. There are three general types of circuits.

(1) Fixed Circuit. This is a circuit in which apparatus of an immovable type (fixed into the ground) is used.

(2) Movable Circuit. This circuit consists of individual exercise apparatus which is portable and can be moved to and from the training area.

(3) Simplified Circuit. This circuit requires no equipment or apparatus.

b. Formation. The exercises are done at will, but rapid, steady, and continuous work is required of all. Each Marine's nervous and muscular system reacts differently to timed vigorous exercises. His performance should be measured on how many movements per exercise he can complete as an individual. For example, one Marine may be able to complete 5 movements, while another may be able to complete 20, and yet each is receiving the maximum benefit. All three circuits contained in this chapter are designed for platoon-sized groups. Expansion beyond this capacity requires a large amount of equipment, as each Marine in the fixed and movable types of circuits must have an item of equipment available for exercise at each station. A group larger than a platoon could be exercised through use of the simplified type of circuit; however, the group would be unwieldy and control could be a problem.

c. Place in the Program. All circuits illustrated can be completed in a 15-minute period. This feature allows the exercise of a platoonor smaller group on the circuit for a single 15-minute period, or the scheduling of the circuit as a 15-minute period within a longer period. A circuit can thus be utilized within a rotating activity system of scheduling. Choice of a circuit by the unit depends upon area, facilities, and other local factors; however, there is a circuit for every need.

3702. FIXED STRENGTH CIRCUIT

a. Description and Objective. The strength circuit is an arrangement of various types of exercise apparatus which are fixed in position. (See fig. 3-16.)Seven basic exercises are used and each exercise requires an apparatus. All apparatus of one type are positioned together to constitute a station. Each station will accommodate 10 Marines. The circuit is designed to be accomplished in 15 minutes when conducted on a time-rotation basis, as normally executed for unit training. For individual use, a Marine may complete the circuit by executing a specific number of repetitions for each exercise. In this case, the time required for completion of the circuit would vary slightly depending on the number of repetitions the Marine completed. The objective of this circuit is to provide a series of exercises which will improve and maintain the strength of the body's major muscle groups.

Figure 3-16 is available in the illustration section at the end of the book.

b. Starting Level. Marines must be thoroughly warmed up prior to participating in the circuit system. If Marines have not engaged in vigorous exercise immediately prior to starting the circuit, then an instructor should execute the following warm up exercises. These exercises should be conducted in the normal formation for set drills. Seven repetitions of each exercise will normally provide sufficient warm up. The exercises are-

(1) High jumper, Conditioning Drill 1.

(2) Bend and reach, Conditioning Drill 1.

(3) Squat bender, Conditioning Drill 1.

c. Progression. The instructor can adjust the circuit system through several methods to accommodate participants of varying physical ability. One method of adjusting an exercise is to change the method of executing the leg lift or to select a heavier weight for the barbell curls. When it becomes apparent during a unit program that the overall fitness of the group has Improved, then the exercise may be made more strenuous in two additional ways. First, the instructor can have one or more of the supplementary stations added. Second, the time spent exercising at each station can be increased in 5-second increments to a maximum of 60 seconds. Also, as Marines learn the circuit, the 45-second period for movement and instruction between stations can be eliminated, thus allowing only enough time to change stations.

d. Starting Positions. The Marine starts on any station, exercises steadily for a certain period (45 seconds initially), then moves on command to the next type of apparatus where he again exercises steadily for an equal period. The Marine continues until the required exercise is completed at each station. If it is desired to expand the number of stations in the circuit to accommodate more participants at one time, the instructor can provide four additional supplementary exercises, requiring no equipment. This will increase the amount of time required to complete the circuit.

e. Leadership. Close control of all Marines is necessary to ensure that a minimum amount of time is spent in moving them to their initial stations and in moving them between stations. One instructor can control the activity on the strength circuit. A stop watch or wrist watch with a second hand is required. When Marines arrive at the strength circuit, the instructor will form them for exercise and conduct the warm up drill. The group is then reassembled and formed into a number of files equal to the number of stations being used in the circuit. Each file is then directed to a station. As soon as all participants have reached an exercise position at a station, the instructor gives the command READY, GO. After 45 seconds of exercise, the instructor gives the command STOP, CHANGE OVER. The instructor allows Marines 45 seconds for moving to the next station and for preparing for the next exercise before the command EXERCISE is again given. In lieu of verbal commands, a whistle may be used to stop and start the exercises. For large groups, a megaphone or loudspeaker is useful.

f. Individual Conditioning Program. For the Marine working alone on the strength circuit, it would be impractical to time the exercise periods. For individual exercise, the Marine should select a number of repetitions of each exercise to complete, then rotate to the next stations after completing these repetitions. The number of repetitions selected should be at or near the maximum that the Marine is capable of doing without halting for rest.

g. Fixed Circuit Stations

(1) Primary Stations (See fig. 3-16.)

(a) Pullups/Chinups. A horizontal bar placed 8 feet above the ground is necessary for each Marine at this station. The Marine will also need a space on the bar that is 45 inches wide. On the command**EXERCISE**,the Marine grasps the bar with both palms facing either forward or to the rear, arms fully extended, and feet free of the ground and executes the exercise as described in MCO 6100.3_.The exercise is repeated as many times as possible until the command **STOP** is given. Then the Marine moves to the next exercise. If a Marine has done his maximum number of pullups/chinups prior to the command **STOP**, he will remain in the "dead" handing position until the command **STOP** and move to the next station.

(b) Twist Grip. The apparatus is a horizontal bar, free to turn, held between uprights placed 30 inches apart. The bar is 52 inches above the ground. A weight of 20 pounds is attached to the center of the bar by a light rope long enough to permit the weight to rest on the ground. The Marine stands at arm's length from the bar and grasps it with his hands on either side of the rope, palms down, thumbs under the bar. On the command **EXERCISE,** the hands are rotated so that the backs of the hands are rotated away from the body, thus winding the rope on the bar. The elbows are kept straight to ensure that the exercise is performed by the hand and forearm. When the weight is drawn up to the bar, the bar is then rotated in the opposite direction to lower the weight to the ground. This exercise is continued until the command **STOP** is given. The Marine then moves to the next station.

(c) Pulley Weights. The apparatus is a T frame with a system of pulleys that suspends a weight of about 90 pounds. The weight is attached to a light steel cable which has a drawbar attached to the other end. The Marine grasps the drawbar and sits down directly under the bar, legs extended to the front and arms extended overhead. The exercise is executed by pulling the drawbar down behind the head, then extending the arms slowly again until they are fully extended overhead. The exercise is repeated as many times as possible until the command STOP is given. The Marine then will move to the next exercise. Upon completion of the exercise, the weight is lowered slowly to the ground.

(d) Barbell Curls. A barbell is necessary for each Marine at this station. The barbell is constructed of 1 1/4-inch pipe 5 feet long, and two concrete-filled No. 10 cans. Each barbell should weigh about 40 pounds. Variance in the weight of the barbells, up to about55 pounds, will allow appropriate overload to be applied to Marines who are above average in strength or weight. The Marine grasps the bar with the palms forward and assumes a standing position with the barbell held in front of the hips, hands approximately shoulder's width apart. On the command **EXERCISE,** the Marine flexes the elbows and draws the barbell up until it touches the upper chest. The elbows remain at the sides. Breath is inhaled with the upward movement and exhaled as the barbell is lowered to the starting position. The exercise is repeated as many times as possible until the command **STOP** is given. The Marine moves to the next apparatus.

(e) Step-Up. The apparatus is a platform or ledge 18 inches high and of such size. At accommodate 10 Marines. The Marine faces the platform and on the command **EXERCISE,** steps up onto the platform, bringing his trailing foot up beside the leading foot. He then steps back down to the original position, stepping down first with the same foot he initially used in stepping up. After 10 repetitions of the exercise, he changes the order of moving the feet to use the opposite leg for stepping up. Repeat this exercise until the command **STOP** is given. The Marine then moves to the next station.

(f) Leg Lift. The apparatus is a horizontal bar constructed as described in (a). To prevent the body from swaying, a horizontal back support is added 40 inches below the horizontal bar. The arms are kept fully extended. On the command **EXERCISE,** the Marine jumps up, grasps the bar with the palms forward and the back support behind him. The exercise is executed by raising the legs to a horizontal position then slowly lowering them to the vertical position. The Marine does not flex his knees. He does not swing the legs to the rear of a vertical position to gain momentum for raising them in the next repetition of the exercise. The movement is repeated until the command **STOP** is given. The Marine then moves to the next exercise. If unable to raise his legs to a horizontal position without flexing his knees, the Marine flexes his knees and draws them up to the chest, then lowers his legs to the vertical position.

(g) Rope Climb. The rope climb is 20 to 30 feet high with five ropes suspended from a horizontal bar which forms the uppermost part of the framework. To prevent the horizontal bar from sagging and to provide safety, only five ropes are attached to it. There are two frameworks per station. The ropes are 6 feet apart. Any method may be used to climb the rope, and the Marines climb as high as possible. Marines who are proficient should climb therope several times during the time allotted. Inexperienced Marines should be cautioned to take care during descent to avoid rope burns on their hands.

(2) Supplementary Stations. The following exercises are designed to expand the basic circuit by being inserted in specific places within the system. For each supplementary station used, there should be adequate room for 10 Marines to exercise.

(a) Bent Knee Situp or Bottoms-Up. These calisthenics are designed to strengthen the abdominal muscles. These exercises will be inserted between the pullup/chinup and twist grip stations. The primary stomach exercise is the situp. In case of inclement weather or other conditions that make ground contact undesirable, the bottoms up exercise is used. In situps on the command **EXERCISE,** the Marine lies on his back (supine position) with knees flexed and both feet flat on the ground and executes the exercise (minus an assistant) as described in MCO 6100.3_.The Marine then returns to the starting position, repeating the exercise until the command **STOP** is given. The Marine then moves to the next station. In bottoms up on the commandEXERCISE,the Marine assumes the front leaning rest and executes the bottoms-up exercise as described in Conditioning Drill 3. He continues this exercise at a moderate cadence until the command **STOP** is given. The Marine then moves to the next station.

(b) Pushup. This exercise is designed to strengthen the arm- and shoulder-girdle muscles. It should be included between the twist grip and pulley weight stations. Upon the command **EXERCISE,** the Marine executes the pushup as described in Conditioning Drill 1. He continues this exercise at a moderate cadence until the command **STOP** is given. The Marine then moves to the next station.

(c) Knee Bender. This exercise is designed to build leg muscles and is included between the pulley weight and barbell curl stations. On the command **EXERCISE,** the Marine executes the knee bender as described in Conditioning Drill 3.He continues this exercise at a moderate cadence until the command **STOP** is given. The Marine then moves to the next station.

(d) Trunk Twister. This exercise strengthens the major muscles of the trunk and is included between the step-up and pullup/chinup stations. On the command EXERCISE, the Marine executes the trunk twister as described in Conditioning Drill 1. He continues this exercise at a moderate cadence until the command **STOP** is given. The Marine then moves to the next station.

3703. MOVEABLE STRENGTH CIRCUIT

a. Description and Objective. The exercises in this circuit are progressive and the course is planned to gain and hold the interest of the participating groups. The circuit consists of a series of stations, with each station designed to develop a particular group of muscles. Along with muscular development, correct posture and deep rhythmic breathing should be stressed on this circuit at all times.(See figs. 3-17 and 3-18.)

Figure 3-17 is available in the illustration section at the end of the book.

b. Equipment. The equipment is set up in files. Six files of 8 stations will accommodate a platoon

c. Formation. The platoon marches to the area where the equipment is positioned and forms a file within each lane of stations, covering down on a piece of equipment. Movements are made on the double, the important factor being that no time is wasted in getting to work.

d. Progression. Initially 40 to 45 seconds per station is adequate. As individuals become stronger the time should be increased in 5-second increments until a minute to a minute and a half is reached.

e. Leadership

(1) The leader stands in front of the barbell station and controls the rotation from this position. The leader supervises the entire group, with the assistance of several instructors who move about in the platoon correcting and encouraging the Marines.

(2) The leader starts each group but does not count cadence nor lead them through the exercises. Each individual exercises rapidly but individually.

(3) As each Marine finishes his repetitions with the barbell, he places the barbells on the ground. The leader calls, READY, followed by the command, FALL OUT ONE. All will double-time to the station directly in front of them, while the individuals on the barbell stations will do an about face to the rear station in their lane.

f. Movable Circuit Stations. The Marine can obtain best results on the movable circuit if the exercises on the various stations are given in the following manner:

(1) Station 1—The Barbell. (See fig. 3-17, 1.)The exercises at this station stress proper posture; deep, rhythmic breathing; and development of the muscles of the arms, shoulders, and upper body. An instructor teaches the proper methods of lifting before the exercises begin. The methods are to lift with the legs, to keep the back straight, and to merely grip with the hands. Two recommended exercises are given below. The instructor will use only one exercise per period .He may use either exercise. At successive periods, the other exercise is used.

(a) **Exercise 1**--Two-Hands Military Press. (See fig. 3-18, 1.)Grasp the barbell with both hands, knuckles up at shoulder width, and lift to the chest. Steadily press to arm's length overhead; lower to the chest resisting weight all the way. Inhale as the weight is pressed up and exhale as the weight is brought down.

(b) Exercise 2--Two-Hands Regular Curl. (See fig. 3-18, 1.) Lift the weight to the waist, with the palms of the hands out, heels together, stomach in, chest lifted and arched, shoulders back, elbows in close to the sides; inhale deeply and curl the weight to the shoulders, using the arms only, at the same time keeping the elbows close to the sides; exhale Emphasize posture and the use of the arms only. This is a very valuable exercise for the development of the biceps and the grip and should be repeated from 8 to 16 times, depending on the ability of the Marine.

(2) Station 2--The Jump Rope. (See fig. 3-18, 2.) This exercise develops strength and agility in the legs and stamina of the whole body. It makes the Marine agile on his feet and increases his footwork efficiency and timing. Each Marine should progress until able to jump rope at least 3 minutes at top speed.

(3) Station 3--The Twist Grip. (See fig. 3-18, 3.) The twist grip is an excellent exercise for the hands and forearms, and adds greatly to the Marine's ability in hand-to-hand combat. The handle is gripped and twisted, winding the rope until the weighted can is level with the height of the hands, which are held horizontal. The weight is lowered in the same manner; the Marine resists the weight all the way, occasionally stopping the twisting motion and alternately removing first one hand, then the other, from the handle. A variation of the above exercise is to wind the handle with the palms up and the arms bent and with the elbows held close in to the sides. Each Marine should maintain a good posture and keep the stomach muscles taut throughout this exercise.

(4) Station 4--The Incline Plane. (See fig. 3-18, 4,)The use of the incline plane is a very strenuous exercise and well designed for the development of the abdomen. Although 6 to 10 repetitions are sufficient for the beginners, more can be added as ability increases. Marines with hernias or recent operations will be excused from participation at this station.

(5) Station 5--The War Club. (See fig. 3-18, 5.) The war club is a simple and effective means of exercising the principal muscle groups of the body, especially those of the trunk, back, and shoulders. To gain the maximum benefit from this exercise, the Marine must keep both feet flat on the ground at all times. Throughout the exercise period, the weight is swung from arm's length as follows:

(a) As in chopping wood, first on one side, then on the other.

(b) As a batter warming up with a number of bats.

(c) In large circles, first with one hand and then with the other.

(6) Station 6--The Bicycle Ride. (See fig. 3-18, 6.) The bicycle ride is well suited to exercising many of the muscle groups of the body, particularly those of the abdomen. Vary the speed of the exercise, but keep everyone "riding" the entire period. A variation exercise may be performed by placing the legs together, raising them slowly to a height about 2 feet from the ground, and then lowering them slowly to the ground.

(7) Station 7--The Step-up. (See fig. 3-18, 7.) The step-up exercises the legs. The step-up is performed by initially stepping up with the left foot, followed by the right, then stepping down with the left foot followed by the right. Continue for 20 seconds, then change to the right foot as the lead foot for 20 seconds.

(8) Station 8--The Isometric Pull. (See fig. 3-18, 8.) Two trainees work at this station with a cable pull and perform the following exercises:

(a) Initially start with one Marine in the supine position and one Marine sitting. The sitting Marine lowers the upper body to the ground and pulls the partner up to the sitting position. The partner then performs this same action and this is continued for 20 seconds at a rapid rate.

(b) During the last 20 seconds, the same action takes place but, in this case, the Marine in the supine position resists the pull of the partner for approximately 5 seconds before allowing to be pulled up into the sitting position.

3704. CIRCUIT-INTERVAL TABLE

a. Description and Objective. The circuit-interval table is designed to develop strength and endurance within a short period of time, with no equipment requirement, through a rapid and vigorous routine of exercise. Fifteen minutes is an adequate period to execute all exercises and to secure a vigorous workout with the circuit-interval principle.

Figure 3-18 is available in the illustration section at the end of the book.

b. Formation and Starting Positions

(1) A leader forms platoons or smaller groups in an oval or circular formation with 3- to 5-yard intervals between Marines. The Marines face to the right and move forward at quick time and then double time. (See fig. 3-19, A.)After running several platoon circle laps, the leader calls out the name of an arm and shoulder exercise from the list below, orders quick time and commands, for example, PUSHUPS. On this command, all Marines immediately hit the ground and individually and rapidly begin doing pushups. No cadence is counted. (See fig. 3-19, B.) After 30 seconds of exercise, the leader commands. **QN YOUR FEET, FORWARD, MARCH** The platoon resumes the quick time cadence and the leader, when ready, gives the necessary commands for double time. The double time is continued for one or more laps and the leader calls out the name of the next exercise and the process is repeated. This continues, with running between each exercise, until every body part has been exercised.

(2) The instructor controls the running and quick time to observe the effects of the exercise upon the Marines. Cadence, step, and precision are not important to the objective and the instructor should not use them. What is important is speed and the instructor should stress this. After the exercise period is started, the Marines do not stop. This circuit method emphasizes stress and recovery, the recovery occurring during the quick time periods.

c. Activities. The leader can use the following exercises. These exercises can be repeated if necessary, during a second round. The leader can use other calisthenic exercises in the circuit-interval table.

(1) Arms and shoulders-push-ups.
(2) Stomach situps.
(3) Back-squat thrusts.
(4) Legs-bicycle (on back).

d. Progression. lhe progress is controlled by the leader, who must pace the running, quick-time movement, and exercise in such a way that Marines will receive a vigorous workout yet be able to participate throughout the 15-minute period. Marines who are in the initial stages of physical condition will not be able to double time or exercise as long as those who are better conditioned. The idea is to set a pace which can be increased during each workout, thus progressing gradually to a higher level of physical fitness.

e. Leadership. The platoon leader, platoon sergeant, or section leader can lead the group. The leader must execute the exercise with the unit to feel the effects and thereby adjust the pace.

f. Place in the Program. This activity may be scheduled whenever a short period of time is available. The only requirement is that enough space, indoors or out, be available to form the circle.

Figure 3-19 is available in the illustration section at the end of the book.

Section VIII. Basic Physical Skills and Obstacle Courses

3801. GENERAL

The purpose of this section is to list the basic military skills and the methods for their development. Many of these skills are best developed by obstacle courses but other drills are also discussed. The section explains types of obstacle courses, details of construction, and methods of negotiating the various obstacles.

3802. BASIC PHYSICAL SKILLS

a. Objective. The objective is to develop proficiency in the various military physical skills which are essential to personal safety and effective combat operations. In travel by foot over rugged terrain and in the execution of combat duties, Marines must be trained to perform basic skills, such as running, jumping, climbing, and carrying. During training, Marines will develop agility and coordination in these skills .Fast and skillful execution of these skills may mean the difference between success and failure on the battlefield.

b. Place in the Program. These skills are practiced throughout the entire program in many different activities. Many of these skills are best practiced on obstacle courses.(See par. 3803.)

c. Description. These basic skills are the minimum skills required by all Marines. The basic skills are as follows:

(1) Running. Running is used to strengthen the legs and develop cardiovascular endurance. Marines should be exposed to running in various situations: on roads, over rough ground, up and down hills, cross-country, and running over low obstacles.

(2) Jumping. In broad jumping, the take-off foot is planted firmly and the spring comes from the extension of this leg as the other leg reaches for the far side of a ditch or similar obstacle. (See fig. 3-20.) The arms are forcibly raised forward and upward to assist in propelling the body up and forward. Landing may be on one or both feet depending upon the length of the jump. In vertical jumping downward from a height, the jumper should aim his feet at the desired landing spot and he should also jump with the knees slightly bent and feet together, with the trunk inclined slightly forward. As the feet touch the ground, the shock is absorbed by bending the knees into a full squatting position. If the height is too great or the ground too hard to absorb the shock, then the jumper should forward roll or side roll thus eliminating some of the momentum.

Figure 3-20 is available in the illustration section at the end of the book.

(3) Dodging In combat situations, it is often times necessary to change directions quickly. To execute this movement while running, a lead foot is firmly planted, left foot if the direction is to the right and right foot if the direction is to the left. The opposite foot is moved toward the new direction. The knees are slightly flexed during the movement and the center of gravity is low and balanced. (See fig. 3-21.) At the time of the change of direction, the head and trunk are turned quickly in the new direction.

Figure 3-21 is available in the illustration section at the end of the book.

(4) Climbing and Surmounting. All Marines should know how to climb and surmount various types of obstacles. (See fig. 3-22.)

Figure 3-22 is available in the illustration section at the end of the book.

(a) Vertical Climbing, as in Climbing a Rope or Pole. The hands grasp the rope or pole overhead with the palms toward the face. Gripping the object, the body is pulled upward with the arms and shoulders, assisted by the feet which grip the object and assist by pushing downward. If shoulder girdle strength and body coordination are not adequate to permit alternating the hands, the arms act together in pulling upward.

(b) Climbing as in Surmounting a Wall. Ingoing over a wall, the body should be kept as close to the top as possible, since in combat operations it is important to offer as small a target as possible to the enemy. If an individual climbs a wall while carrying a rifle, both hands should be freed by slinging the rifle over the back. There are two methods commonly used for surmounting a wall of moderate height, but only one for dropping from it.The methods are as follows:

1 Run, Jump, and Vault. Approach the wall at a run, jump forward and upward at the wall and place one foot against it as high up as possible. Use the foot in contact with the wall to help push the body upward while grasping the top of the wall with the hands. Pull the body up with the arms, assisted by pressure of the foot against the wall and swing the legs over, propelling the body weight over the wall.

2 Hook and Swing. Approach the wall at a run and jump forward and upward. Hook one elbow over the wall, locking the arm in place by pulling up until the top of the wall is underneath the armpit. Grasp the top of the wall with the other hand. Draw the leg which is closer to the wall up as far toward the top as possible. Then swing the other leg over the top of the wall.The body is then carried over with a rolling motion. A variation of this leg action can be used by Marines who are unable to draw up the leg as described. While hanging with both legs fully extended, start a swinging motion with the legs together. When the legs have enough momentum, swing the outside leg over the top of the wall with a vigorous kick, then follow with the body.

3 **Dropping**. All drops from a height are executed in the same manner regardless of the method used to gain the top. One hand is placed against the far side of the wall while the other hand grasps the top. From this position, the body is rolled over the wall and vaulted away from it with the legs swinging clear. As the body passes over the wall and drops, it should at all times face the wall. This will keep the rifle and other equipment clear. Break the fall by retaining a grasp on the top of the wall as long as possible.

(c) Climbing Ladders and Cargo Nets. Rope ladders, stationary vertical ladders, and cargo nets employ the same general technique. The important element is to grasp the side supports firmly in the hands about shoulder height and place the feet on a rung which would cause the body to be fully extended. In movement upward, one hand is moved upward and a new grasp is secured and, at the same time, the opposite leg moves up a rung. As the knee straightens, the body is elevated. This process is repeated using the opposite arm and leg. Alternation continues in this manner until the climber reaches the objective.

(5) Traversing Horizontal Objects. The traversing of horizontal objects puts heavy stress on the arms and shoulder girdle area as the feet are usually suspended in the air with all of the body weight on the arms and shoulders.(See fig. 3-23.)

(a) Traversing Horizontal Ropes or Pipes. The hands grasp the horizontal support overhead with the palms facing. To propel the body forward, one hand is released and moved forward to secure a new grasp. At the same time, the opposite side of the body is swung forward (some people are able to "walk" in the air, keeping the body to the front and moving the legs in time with the arms as in walking on the ground). The other hand is then released and moved forward; this alternation is continued until the objective is reached.

Figure 3-23 is available in the illustration section at the end of the book.

(b) Traversing Horizontal Ladders. In this situation, the movement is the same as used in traversing a rope or pipe. The hands, however, are placed on the rung with the palms away from the face. Other than this difference the technique is the same.

(6) Crawling. Crawling in combat situations is an often used skill. Crawling may be high or low.(See fig. 3-24.)

Figure 3-24 is available in the illustration section at the end of the book.

(a) High Crawl.

In the high crawl, the Marine moves on hands and knees, moving one hand and the opposite knee and then continuing to move the hands in alternation with the opposite knee following the companion hand.

(b) Low Crawl. The Marine is in the prone position usually with the forearms and palms of the hand on the ground. He propels forward by bending the knee of one leg and pushing with the inside edge of the shoe. At the same time, the opposite arm moves forward and pulls to the rear. The body remains low and movement is continued by bending the opposite knee and pushing, and at the same time sliding the opposite arm forward and pulling. Alternation of hands and legs continues until the objective is reached.

(7) Throwing. Throwing may be executed from the kneeling or standing positions. The object to be thrown is held in the hand, and the throwing arm is bent at the elbow; the hand is then moved to the rear until the hand is behind the ear. The body is turned so that the lead foot and balance arm on the side toward the target are pointing at the target. (See fig. 3-25.) The balance arm is used to sight over and align the throwing hand and the target. When properly aligned, the elbow is moved rapidly forward until it is at a point just in front of the body where the arm is straightened and the wrist snapped. This whip motion propels the object to the target. Underhand throws secure momentum by the thrower bending his knees and swinging the throwing arm to the rear. As the knees are straightened, the arm is forcefully swung forward from the shoulder and the object released.

Figure 3-25 is available in the illustration section at the end of the book.

(8) Vaulting. Vaulting is employed to overcome low barriers or fences. (See fig. 3-26.)The object to be surmounted is approached at an angle. The hand on the side next to the obstacle is placed on the top of the obstacle and, with a straight arm, the body weight is pushed upward. At the same time, the leg on the side next to the obstacle is thrown upward and over the top followed by the other leg (side approach). In landing, the weight comes down on the leading leg first followed by regaining the balance on both legs. The free arm serves as a balance. A direct (front) approach can be used at which time both legs go over the object together.

(9) Man Carrying. There are three basic individual means of carrying personnel in combat situations and one of these methods may be used in carrying objects.

Figure 3-26 is available in the illustration section at the end of the book.

(a) Fireman's Carry."A" stands sideways in front of "B", "A" bends his knees and leans forward, placing one arm through "B's" crotch, grasps the wrist of "B's" arm, which is hanging over the shoulder, and then "A" runs forward.

(b) Saddle-Back Carry. With his back toward "B", "A" stands in front of "B". "B" mounts "A's" hips and clasps his arms in front of "A's" chest. "A" grasps "B's" thighs.

(c) Single-Shoulder Carry. "A" stands facing "B"."A" assumes a semi-squatting position. "B" leans forward until "B" lies across "A's" right shoulder. "A" clasps his arms around "B's" legs and straightens up, lifting "B" from the ground. "A" then runs forward. This method may also be used to carry heavy objects.

(10) Balancing. Balancing the body while walking or running on a narrow object when crossing obstacles is a skill which requires practice and confidence. Balance is required in negotiating a log placed across a stream, in crossing a narrow beam or rail, and in similar situations.(See fig. 3-27.) To perform this skill, place the feet on the object to be crossed, hold the arms to the sides at shoulder level, and fix the eyes on the object approximately 5 yards in front of the feet. Generally, it is not a good practice to look down at the feet. Walk the beam by placing first one foot and then the other in the center of the beam, thereby moving forward, using the arms to aid in maintaining balance.

Figure 3-27 is available in the illustration section at the end of the book.

(11) Falling. Injury can be avoided if Marines are taught to fall properly. They should know how to use the body momentum to their advantage during a fall rather than to try resisting that force. (See fig. 3-28.) If enough force is present, such as occurs during a fall while running or in jumping downward from a height, individuals can extend their hands to catch the weight. At the same time, duck the head and roll forward onto the feet. The key to falling without injury from the standing position is relaxation and rolling on the outside of the leg, hip, and buttocks to take the brunt of the fall.

Figure 3-28 is available in the illustration section at the end of the book.

3803. OBSTACLE COURSES

a. General. Obstacle courses are a valuable part of physical readiness training. The challenge presented by the obstacles assists in developing and testing the basic physical skills. In many combat situations, success will depend upon the Marine's ability to perform one or more of these skills, often while carrying field equipment and when fatigued. In this section, two types of obstacle courses will be discussed-Conditioning Obstacle Course and Confidence Obstacle Course.

b. Course Safety Precautions. Commanders and course instructors should take certain precautions to prevent injury on obstacle courses. A few of the precautions are:
(1) Inspect the course for faulty construction of obstacles, protruding nails, rotten logs, condition of the landing pits, and other hazards to safety.
(2) Conduct warm up exercises before the unit runs the course.
(3) Explain and demonstrate the correct techniques for negotiating all the obstacles before allowing the Marines to try them.
(4) Give Marines at least two weeks of conditioning exercises before scheduling the obstacle and/or confidence courses.
(5) Ensure that negotiation of the higher and more difficult obstacles are under the supervision of an instructor.
(6) Do not permit individuals who have neither practiced the basic skills nor run the conditioning obstacle course to participate in the confidence obstacle course.
(7) Weather conditions may cause footing or handhold surfaces to be slippery. If such is the case, postpone training on the course.

3804.CONDITIONING OBSTACLE COURSES

a. Description and Objective. The Conditioning Obstacle Course is commonly known as the Obstacle Course. This course consists of fairly low obstacles which are designed to be negotiated quickly. The obstacles serve to test various basic skills, and running the course is a test of the Marine's physical condition. After receiving instruction and an opportunity to practice the skills, Marines run the course against time.

b. Area and Equipment
(1) Complete standardization of obstacle courses should not be attempted since topographical conditions always vary. Commanders should use ingenuity in constructing a course, making good use of streams, hills, trees, rocks, and other natural obstacles. Since the course is eventually run at high speed, it should not be dangerous.

(2) The course should be wide enough for six or eight men to run simultaneously, encouraging competition. The lanes for the first several obstacles should be wider and the obstacles themselves easier than those that follow. This avoids congestion until the contestants scatter out over the course. The last two or three obstacles should not be too difficult and should not involve high climbing. This prevents injuries and falls resulting from fatigue.

(3) The total distance of the course should range from 300 to 450 yards and include from 15 to 25 obstacles. Normally the obstacles should be 20 to 30 yards apart and arranged so that those which exercise the same groups of muscles are separated.

(4) The obstacles should be substantially built. Peeled logs, 6 to 8 inches in diameter, are ideal for many of the obstacles. Sharp points and corners should be eliminated. Landing pits for jumps or vaults should be filled with sand or sawdust to prevent injuries.

(5) The course should be constructed and marked so that it is not possible to sidestep or detour obstacles. However, it is desirable to provide alternate obstacles of varying degrees of difficulty.

(6) The course should be in the shape of a horseshoe or figure eight so that the finish is close to the start and signs should be placed to indicate the course route.

c. Leadership. Before Marines run an obstacle course, they should be instructed in the proper technique of negotiating each obstacle. In each case this technique should be explained and demonstrated in detail, with emphasis on avoiding injury. Every individual should be given an opportunity to practice on each obstacle until he becomes reasonably proficient at negotiating it. Before the course is run against time, it is advisable for individuals to make several runs at a slower pace. During such practice or trial runs, the instructor should observe the performances and make appropriate corrections. Marines should never be permitted to run the course for time until they have practiced on all obstacles. The best method of timing the runners is to have the timer stand at the finish and call out the minutes and seconds as each individual finishes. If several watches are available, each wave may be timed separately. If only one watch is available, the different waves should be started at regular intervals, such as every 30 seconds. If an individual fails to negotiate an obstacle, a previously determined penalty should be exacted.

d.Types of Obstacles

(1) Jumping-Type Obstacles. These obstacles may be ditches which are cleared with one leap, trenches which the individuals can jump into, heights which require jumping downward, or hurdles which an individual can jump over. (See fig. 3-29.)

Figure 3-29 is available in the illustration section at the end of the book.

(2) Dodging-Type Obstacles. Obstacles of this type are usually mazes consisting of posts set in the ground at irregular intervals. (See fig. 3-30.)The intervals between posts should be rather narrow so that the Marines must pick their way carefully through and around them. Lane guides may be constructed to guide the Marines to dodge and change direction. Obstacles may be put into a maze pattern to cause the Marines to change direction.

Figure 3-30 is available in the illustration section at the end of the book.

(3) Vertical Climbing and Surmounting Type Obstacles. These obstacles may be climbing ropes, either plain or knotted and 1-5 inches in diameter; cargo nets or walls 7 or 8 feet high; or vertical poles 6 to 8 inches in diameter and 38 feet high.(Seefig. 3-31.)

Figure 3-31 is available in the illustration section at the end of the book.

(4) Horizontal Traversing-Type Obstacles. Horizontal obstacles may be pipes, beams, ladders, or ropes. (See fig. 3-32.)

(5) Crawling Type Obstacles. Obstacles which require crawling may be constructed of large pipe sections, low rails, and wire. (See fig. 3-33.)

Figure 3-32 is available in the illustration section at the end of the book.

(6) Vaulting-Type Obstacles. Obstacles of 3 to 3.5 feet in height such as low walls or fences may be used as a vaulting obstacle.(See fig. 3-34.)

(7)Balancing-Type Obstacles. Beams, logs, and planks may be used as balancing-type obstacles.(See fig. 3-35.)These items may be used to span water obstacles and dry ditches, or raised off the ground somewhat to simulate natural depressions.

Figure 3-33 is available in the illustration section at the end of the book.

3805.CONFIDENCE OBSTACLE COURSES

a. Description and Objective. The Confidence Obstacle Course is commonly known as the Confidence Course. This course is composed of higher and more difficult obstacles than those used in the conditioning course. The confidence obstacle course is designed to give Marines confidence in their mental and physical capacities and to cultivate their spirit of daring. They are encouraged but not compelled to negotiate this course and the course is not run against time. The negotiation of a confidence course, however, is strenuous enough to be an excellent physical conditioner. Marines should NEVER attempt to take the obstacles at high speed and should not compete for speed. The obstacles vary from fairly easy to extremely difficult. Some are of considerable height to accustom Marines to climbing such heights without fear. Considerable emphasis is placed on obstacles that train and test an individual's balance.

Figure 3-34 is available in the illustration section at the end of the book.

b. Area and Equipment

(1) The confidence course accommodates four platoons, one platoon at each group of six obstacles. The course should be made up of about 24 obstacles, numbered and marked as follows:! to 6, white numbers on red background; 7 to 12, black numbers on a white background;

13 to 18, white numbers on a blue background; and 19 to 24, white numbers on a black background.

(2)A few simple pieces of equipment should be provided for individuals who do not have the strength, courage, or ability to negotiate the obstacles.

Figure 3-35 is available in the illustration section at the end of the book.

c. Formation. The obstacles should be divided into groups of six, and each group is designated by a different color. Each platoon starts at a different color. Individuals are separated into groups of 8 to 12 at each obstacle. At the starting signal from the company commander, they proceed numerically through. Anyone may skip an obstacle who is afraid to try. Individuals proceed from obstacle to obstacle until time is called, then assemble as ordered.

d. Leadership. If the Marines are new to the confidence course, an instructor will demonstrate or will give a brief orientation at each obstacle, including an explanation and demonstration of a method of negotiating it. Marines are encouraged to try the various obstacles, but they are not compelled to do so. No compulsion is to be used. The manner of negotiating any obstacle is left to the discretion of the Marine. However, the instructor assists anyone who experiences difficulty. Instructors must supervise closely at all times to prevent injuries, as some of the obstacles are quite high. Also, some of the obstacles should not be used when slippery or wet. The example of instructors and especially selected demonstrators will serve to inspire the individuals to greater effort.

e. Negotiating the Obstacles. Although personnel need not conform to any one method of negotiating the obstacles, there should be some uniformity in the approach to them. A general method of negotiating the obstacles is indicated below.

(1) Red Group This group contains the first six obstacles. (See fig. 3-36.)

(a) The Belly Buster Individuals may vault, jump, or climb over. Warn them that the log is not stationary.

(b) Reverse Climb. Climb the reverse incline and go down the other side to the ground.

(c) The Weaver. Move from one end of the obstacle to the other by weaving the body under one bar and over the next.

(d) Hip-Hip. Step over each bar, either alternating legs or using same lead leg each time.

(e) Balancing Logs. Step up on log, retaining the balance, walk or run along log.

(f) Island Hopper. Jump from one log to another until the obstacle is negotiated.

(2)White Group. This group is composed of the second six obstacles. (See fig. 3-37.)

(a) Tough Nut. Step over each X in the lane.

(b) Slide for Life. Climb the tower, grasp the rope firmly, and swing the legs upward. Hold the rope with the legs to distribute the weight between them and the arms. Braking the slide with the feet and legs, proceed down the rope. Warn Marines that there is danger of getting rope burns on their hands. When the rope is slippery or wet, this can be a dangerous obstacle.

(c) Low Belly Over. .Mount the low log and jump onto the high log, both arms grasping over the top of the log, the stomach area in contact with it. Swing the legs over the log and lower the body to the ground.

(d) Belly Crawl Move forward under the wire, belly down, to the end of the obstacle.

(e) The Dirty Name Mount the low log and jump to or reach the higher logs in succession, then jump or drop to the ground. Warn the Marines about the height of the final log.

(f) The Tarzan Mount the lower log and walk the length of it and each successive, higher log until reaching the horizontal ladder. Grasp two rungs of the ladder and swing the body into the air. Negotiate the length of the ladder by releasing one hand at a time and swing forward, grasping a more distant rung.

(3) Blue Group. This group is formed by the third group of six obstacles. (See fig. 3-38.)

(a) High Step over. Step over each log, alternating the lead foot or using the same lead foot.

(b) Swinger. Climb onto the swinging log and over to the ground on the opposite side.

(c) Low Wire. Move under the wire on the back, using the hands to raise the wire to clear the body.

(d) Swing, Stop and Jump. Gain momentum with a short run, grasp the rope, and swing the body forward to the top of the wall. Release the rope while standing on the wall and jump to the ground.

(e) Six Vaults. Vault over the logs, using one or both hands.

(f) Easy Balancer. Walk up one inclined log and down the one on the other side to the ground.

(4) Black Group. The last group is formed by the final six obstacles. (See fig. 3- 39.)

(a) Incline Wall. Approach the underside of the wall, jump up and grasp the top and pull the body up and over. Slide or jump down the incline to the ground.

(b) Skyscraper. Jump or climb to the first floor, climb up the corner posts or assist each other to any desired floor. Descend to the ground in any desired manner.

(c) Jump and Land. Climb up the ladder to the platform and jump to the ground.

(d) Confidence Climb. Climb the inclined ladder to the vertical ladder. Go to the top of the vertical ladder, then down the other side to the ground.

(e) Belly Robber. Step on the lower log and assume the prone position on the horizontal logs. Crawl over the logs to the opposite end of the obstacle.

(f) The Tough One Climb the rope or pole on the higher end of the obstacle, then go down the ladder and across the log platform. Climb over or between the logs at the end and go down the rope or pole to the ground. Vault over the final log.

Figure 3-36 is available in the illustration section at the end of the book.

Figure 3-37 is available in the illustration section at the end of the book.

Figure 3-38 is available in the illustration section at the end of the book.

Figure 3-39 is available in the illustration section at the end of the book.

Section IX Individual Exercise Programs

3901.GENERAL

Marines at times will **be** stationed on independent duty and consequently will be responsible for their own physical fitness program. This section will assist them in understanding the need for exercise and will aid in the planning and execution of an individual exercise program. Exercise activities included are the Bench Conditioner, 6-12 Plan, Weight Training, and Isometric Contraction. Keeping physically fit is a problem that faces every Marine. Even though we are frequently engaged in training that requires some physical effort, in many cases, it is not enough to prepare us to meet the intense physical demands of combat. Attaining a satisfactory level of physical readiness is not an insurmountable objective. Available time appears to be the most difficult obstacle to the development of physical readiness. In most cases, regular physical training programs are centralized, requiring the individual to temporarily leave the work area. The problems involved in setting an hour aside two or three times each are numerous. However, most of us can devote a few minutes each day to physical fitness with little, if any, impact on our daily work schedule, especially if it does not require us to leave our work area.

a. Type of Program. There are many good physical fitness programs available to the individual or group. Regardless of the type or duration, to be effective, the program must contain exercises that are strenuous and are challenging to the individual. Space will not permit the inclusion of all available means of individual exercise. The programs selected for this chapter have met the requirement of minimum space and minimum time.

b. Need to Augment Program. These programs are quite strenuous and will develop a satisfactory level of physical readiness. However, if the Marine desires additional development of endurance, it is recommended that he supplement these programs with a 15- to 30-minute period of wind sprints and double timing on an alternating daily basis.

c. Progressive Training. If Marines are performing duties which require little or no physical activity, they must plan a physical conditioning program that assures a moderate beginning, moderate but steady progression, and sufficient warmup before starting the vigorous exercise. To avoid organic of bodily harm, a Marine activity without adequate warmup. He should conduct conditioning programs on a daily basis over an extended period of time, never on an unduly accelerated or crash basis.

3902.THE BENCH CONDITIONER

a. Description and Objective. The bench conditioning program uses a modified bench to employ both isotonic (moving) and isometric (stationary) exercises are as the nucleus of the program. The exercises are designed to develop strength and endurance in all the major muscle groups of the body. The principles of progression, overload, and balance are employed when the exercises are performed properly.

b. Area and Equipment. The conditioning apparatus can be constructed in any unit motor pool with welding equipment found in most salvage yards.(See figs. 3-40, 3-41.)Additionally, there are available any number of commercially produced apparatus available. The important thing is that in utilizing this equipment the following exercise routine be adhered to.

c. Starting Level and Progression.. The program consists of two tables, each with 10 exercises. The Marine controls progression by required repetitions or, in some cases, by application of maximum effort. Each table can be completed within 15 minutes.

d. Starting Position. To start the program, the Marine begins with the Table 1 and executes each exercise for the required number of repetitions as indicated. The Marine controls the starting level and progression. When the Marine executes the maximum repetitions for Table I within a 15-minute exercise period, he progresses to Table II. To maintain this level of development, the Marine should also execute the maximum repetitions for Table II within a 15-minute exercise period. The Marine should keep substitution of exercises to a minimum. However, it he completes a full 15 minutes of strenuous exercise and exercises all muscles, then there should be no appreciable difference in the overall development.

e. Bench Conditioning, Table I

(1) Exercise 1: Side-Straddle Hop. This is a two-count warmup exercise done at moderate cadence. The starting position is the position of attention. On count ONE jump slightly into the air, swinging the arms out to the sides and up to a vertical position, hands touching.(See fig. 3-40, A.)At the same time, spread the feet wider than shoulder-width apart. On count TWO, using a slight flexing of the knees and ankles, jump slightly into the air and return to the starting position by swinging the arms back down to the sides. Twenty repetitions of this exercise is the standard dosage throughout the program.

(2) Exercise 2: Hand Walk. Remove the lower horizontal bar. Adjust the upper horizontal bar so that it is high enough to permit a "dead hanging" position, with the feet off the ground.(See fig. 3-40, B.j From the "dead hanging" position, release one hand and drop the arm to the side of the body. Then raise that arm and regrasp the horizontal bar. Release the bar with the other hand and drop that arm to the side. Repeat this as many times as possible.

(3) Exercise 3: Situps. Lie down with the fingers interlocked and placed behind the head. Hook the toes under the foot braces. Raise the trunk and upper body to an upright sitting position, twisting it to the left and then forward and downward until the right elbow touches the left knee.(See fig. 3-40, C.)Lower the body to the starting position .Sit up again but twist the body to the opposite direction as before, touching the left elbow to the right knee. Again lower the body to the starting position. The starting dosage is 20 situps. The Marine should continue the progression until he has attained 40 situps.

Figure 3-40 is available in the illustration section at the end of the book.

(4) Exercise 4: Double Step-upStarting at one end of the bench, step up onto the bench, and walk across it. Step down from the other end; turn around and repeat the process to return to the starting point.(See fig. 3-40, D.)Each return to the starting point constitutes a repetition. The starting dosage is 20 repetitions. Maximum dosage is 35 repetitions. This exercise should be done at a rapid cadence.

(5) Exercise 5: Isometric Bar Lift. Adjust the lower bar so that it is slightly higher than the beltline. Placing the feet on the footplates at the base of the frame, grasp the lower bar so that the hands are spread shoulder-width apart. Assume a crouched position and lift with maximum effort using the arms, back and legs.(See fig. 3-40, E.)Starting dosage is 4 repetitions of a stress time of 5 seconds followed by a 5-second rest prior to the next repetition. The Marine obtains progression by lengthening stress periods to 6 and later 7 seconds. Do not increase the number of repetitions.

(6) Exercise 6: Knee Lift. Adjust the upper bar to the same height used in Exercise 2.Adjust the lower bar so that is stops rearward movement of the hips when the "dead hanging" position is assumed.(See fig. 3-40, F.) Keeping the arms extended, flex the legs and raise the knees as high as possible. Hold this position for 5 seconds, then return to the starting position. After 2 seconds in the starting position, raise the knees again. Each return to the starting position constitutes one repetition. The dosage is five repetitions. Progression is the same as in Exercise 5.

(7) Exercise 7: Isometric Pull. Adjust the lower horizontal bar to a position where it is slightly higher than the beltline. Grasp the handles and pull outward.(See fig. 3-40, G.)Apply maximum effort and hold for approximately 5 seconds. Relax for 5 seconds between repetitions; perform four repetitions. Moving the body closer to or farther away from the bar will change the stress from the upper arms to the forearms. Progression is the same as in Exercise 5.

(8) Exercise 8: Isometric Compression. Maintain the position as in exercise 7.(See fig. 3-40, H.)Grasping the handles in the same manner, press in **with** maximum effort and hold for approximately 5 seconds. Relax for 5 seconds between repetitions. Progression is the same as in Exercise 5.

(9) Exercise 9: Isometric Press. Remove the lower horizontal bar. Adjust the upper horizontal bar until it is about 6 inches lower than the extended arms can reach. Stepping on the footplates at the bottom of the frame, grasp the bar with both hands and push up.(See fig. 3-40,1.)Keep both the legs and arms slightly flexed and the back straight. Apply maximum effort for 5 seconds then relax for 5 seconds. Complete four repetitions. Progression is the same as in Exercise 5.

(10) Exercise 10: Pushups. Grasping the foot braces with both hands, assume the front leaning rest position.(See fig. 3-40, J.)Keeping the back and legs straight, lower the body until the chest is lower than the **hands,** then return to the starting position. The Marine should complete the maximum possible number of repetitions.

f. Bench Conditioning, Table II. There is no limit on the maximum number of repetitions attainable in Exercises 3, 4, and 6 of Table II. The only limit imposed is that the entire program of 10 exercises outlinted in either table should not exceed 15 minutes.

(1) Exercise 1: Side-Straddle Hop. This is a two-count warm up exercise done at a moderate cadence. The starting position is the position of attention. On count ONE, jump slightly into the air, swinging the arms out to the sides and up to a vertical position, hands touching.(See fig. 3-41, A.)At the same time, spread the feet wider than shoulder-width apart. On count TWO, using a slight flexing of the knees and ankles, jump slightly into the air and return to the starting position by swinging the arms back down to the sides. Twenty repetitions of this exercise is the standard dosage throughout the program.

(2) Exercise 2: Pull-up. Adjust the horizontal bar so that is is high enough to permit a "dead hanging" position with the feet off the ground. Grasp the bar with both hands, palms facing forward. By flexing the arms, raise the body to a position where the chin is higher than the bar.(See fig. 3-41, B.)Then lower the body to the "dead hanging" position. Repeat as many times as possible.

(3) Exercise 3: Bench situps. Sit on the bench and hook the feet under the foot braces. With the fingers interlocked behind the head, lean back until the head touches the floor.(See fig. 3-41, C.)Return to the starting position. The starting dosage is 15 situps.

(4) Exercise 4: Step-Up. Face the bench and step up on it with one foot, bringing the trailing foot up next to the leading foot. Step back down again, leading with the same foot used first in stepping up.(See fig. 3-41, D.)Perlorm half of the total repetitions, then change the sequence of moving the feet to use the other leg in stepping up, and repeat the same amount of exercise. The starting level is a total of 40 step-up. **This exercise should be done at a rapid cadence.**

(5) Exercise 5: Isometric Bar Lift\djust the lower bar so that it is slightly higher than the beltline. Placing the feet on the footplates at the base of the frame, grasp the lower bar so that the hands are spread shoulder-width apart. Assume a crouched position and lift with maximum effort using the arms, back, and legs.(See fig. 3-41, E.)Starting dosage is four repetitions of a stress time of 8 seconds followed by a 5-second rest prior to the next repetition. Progression is obtained by lengthening stress periods to 10 seconds. Do not increase the number of repetitions.

Figure 3-41 is available in the illustration section at the end of the book.

(6)Exercise 6: Leg Lift. Adjust the bars and assume the starting position as shown in figure 3-41, F. Keeping arms and legs extended, raise the legs to a horizontal position and hold in that position for 2 seconds. Then lower the legs slowly to the starting position. Five repetitions is the starting level.

(7) Exercise 7: Isometric Pull. Adjust the lower horizontal bar so that it is slightly higher than the beltline. Grasp the handles and pull outward.(See fig. 3-41, G.)Apply maximum effort and hold for approximately 8 seconds. Relax for 5 seconds between repetitions; perform four repetitions. Moving the body closer to or farther away from the bar will change the stress from the upper arms to the forearms. The Marine obtains progression by lengthening the stress period to 10 seconds.

(8) Exercise 8: Isometric Compression. (See fig. 3-41, H.)Maintain the position as in exercise 7.Grasping the handles in the same manner, press in with maximum effort and hold for approximately 8 seconds. Relax for 5 seconds between repetitions; perform four repetitions. The Marine obtains progression by lengthening the stress period to 10 seconds.

(9) Exercise 9: Isometric Pres. Remove the lower horizontal bar. Adjust the upper horizontal bar until it is about 6 inches lower than the extended arms can reach. Stepping on the footplates at the bottom of the frame, grasp the bar with both hands and push up.(See fig. 3-41,1.)Keep both the legs and arms slightly flexed and the back straight. Apply maximum effort for 8 seconds, then relax for 5 seconds. Complete four repetitions. The Marine obtains progression by lengthening the stress period to 10 seconds.

(10) Exercise 10: Inclined Pushup. Assume the front leaning rest position with the feet on the bench.(See fig. 3-41, J.)Keeping the back and legs straight, lower the body until the nose touches the ground. By extending the arms, raise the body to the starting position. Repeat as many times as possible.

3903. THE 6-12 PLAN

a. Description and objective. The 6-12 plan of physical fitness has been developed to assist in regulating quantity and progression and to provide a convenient set of exercises. This is a basic program and will take 18 weeks to complete if you follow the moderate progression as to the time prescribed for each level of achievement. The time can be shortened as explained below. This plan consists of six basic exercises a day which can be completed in 12 minutes. There are six tables of six exercises each, thus allowing you to progress from table to table. The plan is progressive, fits any age group, contains balance and variety, and applies the principle of overload in a safe and gradual manner. Begin at Table I, Progression Guide, with the number of repetitions as indicated by age.

b. Formation. If just starting an exercise program, do not rush through the first table. Remember, individuals should remain at each level for about a week before moving upward. The time allotment stated for each exercise at the bottom of the tables is a guide; some people may take more and some less time on the individual exercises. At the end of a one-week period, when the individual can comfortable perform the six exercises in 12 minutes, move on to the next level. To a certain degree, the individual must be the judge of his ability to progress form level to level and table to table. If starting with a certain degree of fitness, some of the beginning tables may present little challenge.

c. Starting Level **and Progression.** There are three levels of achievement for each age group, indicated as A, B, and C. Start at the C level for the appropriate age group. At the end of a one-week period, or when the individual can do all exercises at that level within 12 minutes, progress to the B level. At the end of the second week, or when the individual can accomplish that level within 12 minutes, progress to the A level. At the conclusion of the third week or when the individual can achieve the A level within the time limitation, move on to table II.(See Tables I through VI, Progression Guide.)

d. Maintenance Level. Attempt to work through all six tables. If this proves to be too difficult, then maintain exercise at the—

(1) A-level on Table IV, Progression Guide it in the 45 to 49, 50 to 59, or over 60 age group.

(2) A-level on Table V, Progression Guide if in the 17 to 29, 30 to 39, or 40 to 44 age group.

e. Precautions. To achieve the maximum benefit, perform each exercise exactly as specified. Read the descriptions and study the illustrations. Do not slight the movements. Use a sensible approach and follow these four points as they apply before starting or during your exercise program.

(1) If you have the slightest doubt about your ability to participate in this exercise program, consult a physician.

(2) Stop immediately if you notice unusual breathlessness or chest pain while taking part in these exercises. If these conditions persist, consult a physician.

The following Tables of Progression Guides and Exercises are the progression of the 6-12 Plan program.

Table I. Progression Guide is available in the illustration section at the end of the book.

Table I. Exercises.is available in the illustration section at the end of the book.

Table II. Progression Guide.is available in the illustration section at the end of the book.

Table II. Exercises.is available in the illustration section at the end of the book.

Table III. Progression Guide.is available in the illustration section at the end of the book.

Table III. Exercises.is available in the illustration section at the end of the book.

Table IV. Progression Guide.is available in the illustration section at the end of the book.

Table IV. Exercises.is available in the illustration section at the end of the book.

Table V. Progression Guide.is available in the illustration section at the end of the book.

Table V. Exercises.is available in the illustration section at the end of the book.

Table VI. Progression Guide.is available in the illustration section at the end of the book.

Table VI. Exercises.is available in the illustration section at the end of the book.

(3) Unless you have exercised regularly and know yourself to be in good physical condition, start at table I with the C level appropriate to your age.

(4) If you are out of shape, admit that fact to yourself. Hide your pride; after all, you are in the privacy of your own quarters. Set your goal for the longer, steadier pull toward fitness. Resist the urge to pass over the lower numbered tables to find a table that will test your fitness. You are not trying to test, but rather to develop.

3904.WEIGHT TRAINING

Weight or barbell training should not be confused with the more common types of weight lifting used as a competitive sport. Weight lifting is designed to develop specific muscle groups so that the individual is capable of lifting a large amount during a single lift. In contrast, weight training is the systematic development of all the major muscle groups by the use of calisthenics reinforced with weight to provide resistance.

3905.WEIGHT LIFTING

a. Description and Objective. The weight lifting program is progressive and applies the principle of overload in a safe, gradual manner. The exercises of the table can be completed in 15 minutes. The objective of the exercises is to develop strength and muscular endurance, and muscle tone of the five major muscle groups:legs, arms, back, trunk, and shoulder girdle.(See fig. 3-42.)

b. Warmup. A warm up exercise is important to prepare the body for the more vigorous exercises that are to follow. Ten repetitions of the high jumper exercise are excellent for a warm up period. Also, an exercise which will require fast body movement is needed.To provide such exercise, 3 to 5 minutes of rope skipping is recommended to increase the individual's development potential.

c. Formation. Marines should take care in the completion of these weight lifting exercises. They should ensure that their back is straight during the lifting phase of all exercises.When exercises require assuming the standing position with the weight, they should always grasp the weight while in a squatting position and then rise to a standing position.

d. Progression. Each exercise has a starting number of repetitions and specified pounds of weight. After each 4th or 5th day of exercise, the Marine should increase the repetitions by one until he has reached the maximum of 10 repetitions. At this time, the Marine should increase the weight by 5 pounds. Then he should repeat the process again with the initial number or repetitions.

e. Weight Training

(1) Exercise 1:Squat(See fig. 3-42, A.)Starting level—six repetitions, 50 pounds (commonly called the flatfoot deep knee bend). Place the bar upon the shoulders. Stand with feet about 18 inches apart. Keeping the feet flat, lower the body into the low squat position. Stand erect and repeat. Exhale while lowering into the squat position and inhale while standing. This constitutes one repetition.

(2) Exercise 2:Waist Bender(.See fig. 3-42, B.)Starting level-6 repetitions, 40 pounds. Assume the standing position with the bar across the shoulders and the feet a shoulder-width apart .Bend forward at the waist until the upper body is parallel to the ground; return to the starting position. Each return to the upright position constitutes one repetition.

(3) Exercise 3:Curl (See fig. 3-42, C.)Starting level—6 repetitions, 40 pounds. Grasp the barbell with the palms facing to the rear and assume the standing position with the feet a should-width apart. With the barbell held in front of the hips, flex the elbows and lift the weight until the bar touches the upper chest. Lower the barbell back to the hip level position. Inhale deeply with the upward movement and exhale on the downward movement. Each time that the bar touches the chest will constitute one repetition.

(4) Exercise 4:Side Bender(.See fig. 3-42, D.)Starting level—6 repetitions per side, 40 pounds. Assume the standing position, with the bar across the shoulders, with feet a shoulder-width apart. Bend to the left as far as possible and return to the starting position. Repeat six times and then execute the same procedure to the right for six repetitions.

(5) Exercise 5:Standing Press(See fig. 3-42, E.)Starting level—6 repetitions, 45 pounds. Grasp the bar with the palms facing forward and assume the starting position. Curl the weight to the upper chest position. Inhale deeply and press the bar upward to an overhead position. Exhale while lowering the bar to the chest position. Each time that the bar is pressed upward constitutes one repetition.

(6) Exercise 6:Upward Row. .see fig. 3-42, F.)Starting level—6 repetitions, 40 pounds. Grasp the bar, hands close together, palms to the rear, and assume the standing position. Starting with the bar held in front of the hips, flexing the elbows and the shoulder girdle muscles, lift the bar straight up to an overhead position. Inhale deeply while lifting the bar. Exhale while lowering the bar to the hip position. Each time that the bar returns to the hips constitutes one repetition.

(7) Exercise 7:Shoulder Curl. See fig. 3-42, G.)Starting level—6 repetitions, 25 pounds. Grasp the bar, palms down, and assume the standing position. Keeping the elbows locked, curl the bar, pivoting the arms at the shoulders until the bar is in an overhead position and as far to the rear as possible. Return the bar in the same manner to the hip position. Each time that the bar returns to the hip position constitutes one repetition.

Figure 3-42 is available in the illustration section at the end of the book.

3906.ISOMETRIC EXERCISING

a. Description and Objective. Isometric exercising is the application of maximum effort during an exercise period. The isometric principle is to apply force gradually over a 5- to 10- second period until the maximum effort is applied. Relaxation follows for approximately 5 seconds and then force is applied again. The Marine continues this process at the prescribed level for each exercise for a period of 15 minutes or less. The objective of isometric exercises is to create muscle growth. Isometric exercises are the fastest means of creating muscle growth. They are founded on the fact that a muscle will grow only so fast regardless of the type or duration of the activity. The principle of overload—that the muscles develop commensurate with demand—reinforces that fact. However, isometrics will not develop cardiovascular or muscular endurance. Consequently, Marines who choose to use isometrics in their exercise routine must also include running or some other type of aerobic activity.

b. **Area and Equipment** isometric exercises may be designed to be performed with or without equipment.

c. Use With Other Programs. The isometric exercises presented in this section will provide a variety from which to choose. In addition to the exercises contained in this section, the application of isometric force is used in some of the exercises in section II.

d.Door Frame Exercises. The following exercises are designed for use with a standard door frame found in all offices or barracks.(See fig. 3-43.)

(1) Exercise l:Arm Press. Stand in the doorway with the legs straight, knees locked. Using the arm muscles, press hard upward against the top of the door frame. Repeat for three repetitions applying gradual effort to maximum contraction.

(2) Exercise 2:Leg Press. Stand in the doorway with the hands on the top of the door frame, elbows locked .With knees bent, press hard with the leg muscles .Repeat for three repetitions beginning with a gradual effort and increasing to maximum contraction. A low platform may be necessary to reach the top of the door frame and still maintain a bent knee position.

(3) Exercise 3:Side Press. Extend both arms to the side of the doorway. Palms are shoulder high, facing outward. With both arms, press hard against the sides of the door frame. Repeat for three repetitions. Begin gradually and increase to maximum contraction.

(4) Exercise 4:Lateral Raise. Extend both arms to the sides of the doorway, arms down, palms facing inward. With the back of the hands, press hard against the sides of the door frame .Repeat for three repetitions. Begin with a gradual effort and increase to maximum contraction.

Figure 3-43 is available in the illustration section at the end of the book.

(5) Exercise 5:Neck Press?lace the forehead against the door frame, hands clasped behind the back Using the neck muscles, press hard against the door frame. Repeat for three repetitions, then reverse position so that the back of the head is resting on the door frame. Again do three repetitions .Begin gradually with both exercises and increase to maximum contraction.

(6) Exercise 6:Door Pull Stand facing the edge of the open door and grasp the doorknobs. Pull outward with both arms (if doorknobs are not available grasp the edge of the door).Applying outward pressure, move the body toward and away from the door .Repeat the three repetitions. Begin with gradual effort and increase to maximum contraction.

Chapter 4
COMBAT WATER SURVIVAL
Section I. Marine Corps Water Survival Program

4101.MILITARY SWIMMING

Military swimming emphasizes strokes that result in staying power rather than speed. Marines need to swim easily, quietly, and with adequate vision. Strokes should be used that allow the Marine to carry basic combat equipment and to tow or push a wounded buddy while keeping the face out of the water to allow breathing. The best strokes to use are the side stroke and the breast stroke. Marines must be proficient in floating, drown-proofing, and in using their basic combat equipment to make flotation devices. They must be drilled on using the Kapok and Mae West buoyancy compensators and on emergency egress from helicopters, landing craft, and amphibious assault vehicles.

4102.PROGRAM DEVELOPMENT

Commanders of the Landing Force Training Commands (Atlantic/Pacific) are tasked with developing and maintaining the individual water survival and swimming training program. They are responsible for training all Marine Corps water safety and survival instructors and integrating this training in to their combat readiness training program. All Marines are deployable and as soldiers of the sea must constantly practice the skills necessary to achieve mission accomplishment. Commanders must be imaginative in program development, integrating basic skills development which enhances the total mission accomplishment. An example is teaching waterproofing of the Marine's backpack so that it can be used as a flotation device in streams or rivers. Once learned, this skill would be practiced first in a pool and then in a stream or a river. Routine inspections during deployments would ensure that backpacks could be used as flotation devices.

4103.QUAL STDS AND TRAINING GUIDELINES

The inherent nature of Marine Corps operations and training requires that Marines achieve an ability to survive in water. Water survival and swimming training is designed to reduce fear of water, instill self-confidence, and develop a Marine's ability to survive in water. As a minimum, every Marine should be qualified as swimmer, third-class (S3).This qualification should be met during recruit training or Officer Candidate School. Once qualified, a Marine need not be requalified unless requalification at a specific level (S3 or higher) is required by unit mission, military occupational specialty or duty assignment, or for other reasons determined by the commander. The following classifications and standards apply to Marine Corps water survival and swimming training. Abbreviations will be used for service, medical, and training record entries citing MCO 1510.29A, Individual Water Survival and Swimming Training, as authority.

a. Qualification standards and test procedures are described in paragraph 4104.

b. Swimmer classification and abbreviations are:

(1) Unqualified - UQ.

(2) Swimmer, third-class (S3).Minimum aquatic skill level for all Marines.

(3) Swimmer, second-class (S2)-

(a) Minimum standard for all naval aviators, naval flight officers, and aircrewmen with written waiver from the commanding officer.

(b) Requirements for military occupational specialty 1803 (assault amphibian vehicle officer), 1833 (assault amphibian vehicle crewman), 1302 (engineer officer), 1371 (combat engineer), 1381 (shore party specialist), and 1379 (engineer operations chief).

(4) Swimmer, first-class (S1)-

(a) Naval aviator, naval flight officer, and aircrewman qualification requirement.

(b) Requirements for military occupational specialty 0321 (reconnaissance man), 8652 (reconnaissance man, parachute jump qualified), 8653 (reconnaissance man, self-contained underwater breathing apparatus qualified), 8654 (reconnaissance man, parachute/self-contained underwater breathing apparatus qualified), 9952 (self-contained underwater breathing apparatus Marine (officer/enlisted)), and 9953 (parachutist/self-contained underwater breathing apparatus Marine (officer/enlisted)).

(5) Water Survival Qualified (WSQ).Ultimate water survival/swimming goal for Marines.

c.Instructor and assistant instructor classifications are-

(1) **American Red Cross Advanced Lifesaver.** Individuals who are currently certified as American Red Cross advanced lifesavers may assist qualified instructors during Marine Corps water survival/swimming training and testing. Certification is valid for 3 years.

(2) American Red Cross Water Safety Instruction. Individuals who are currently certified as American Red Cross water safety instructors may conduct swimming training and qualify individuals as 3d, 2d, and 1st class swimmers. Certification is valid for 2 years.

(3) United States Marine Corps Water Safety and Survival Instruction. Those personnel certified as water safety and survival instructors by one of the landing force training commands may conduct Marine Corps water survival and swimming training and American Red Cross swimming and water safety training. These instructors may also participate as assistant instructors during the conduct of landing force training command-sponsored water safety and survival instructor courses. Certification is valid for2 years.

(4) American Red Cross Water Safety Instructor Traine. Individuals qualified who have been certified as instructor trainers by the American Red Cross may train American Red Cross water safety and basic swimming instructors. Certification is valid for 2 years.

(5) United States Marine Corps Water Safety and Survival Instructor Traine. Individuals qualified who are certified as instructor trainers through special training courses conducted jointly by the American Red Cross and the Marine Corps, under the direction of the Commandant of the Marine Corps, may train water safety and survival instructors and American Red Cross water safety and basic swimming instructors. Certification is valid for 2 years.

d. Training Guidelines Instructors engaged in training Marines as swimmers will use the current edition of the American Red Cross swimming and water safety courses instructors' manuals as the authoritative textbooks for swimming procedures, strokes, breaks, and holds.

4104.WATER SURV/SWIM QUAL STD AND TEST PROCS

a. The following qualification standards and test procedures are applicable to Marine Corps water survival and swimming training. The utility uniform without cover, boots/oxfords or socks will be worn for all swimming tests through swimmer, first class. The utility uniform with boots/oxfords will be worn as prescribed for the water survival qualification.

(1) Swimmer, Third-Class (S3) To qualify as a swimmer, third-class, a Marine must enter the water feet first from a minimum height of 5 feet and remain afloat for 5 minutes. During this time the Marine must swim 50 yards using any stroke or combination of strokes. This test should be taken by all Marines as early as possible in their initial training. Those unable to pass this test will be classified as UQ and should be given instruction in fundamental swimming skills. Those who barely meet requirements are swimmers who need help and should receive additional instruction in fundamental swimming skills, whenever possible.

(2) Swimmer, Second-Class (S2.)To qualify, a Marine must enter the water feet first from a minimum height of 10 feet and remain afloat for 10 minutes. The Marine must swim 100 yards and use the three basic survival swimming strokes (side stroke, backstroke, and breast stroke) for a minimum distance of 25 yards each.

(3) Swimmer, First-Class (S1)A prerequisite to qualification as swimmer, first-class, is successful completion of the test for swimmer, second-class. To qualify as a swimmer, first-class, a Marine must be able to do each of the following:

(a) Approach a person of approximately the same size while in the water, demonstrate a release, get that person in a carry position, and tow the "distressed swimmer" 25 yards.

(b) Enter water feet first and immediately swim underwater for 25 yards. Swimmer is to break the surface twice for breathing during this distance at intervals of approximately 25 feet.

(c) Remove trousers in water, inflate for support, and remain motionless for a minimum of 1 minute.

(d) Swim 220 yards using any survival swimming stroke or any combination of survival strokes.

(4) Water Survival Qualification (WSQ)A prerequisite to water survival qualification is successful completion of the test for swimmer, first-class. To be water survival-qualified, a Mannemust:

(a) Enter water by jumping from a height of 10 feet.

(b) Stay afloat in full utility uniform (less cover) for 1 hour. Boots/oxfords will be removed after 5 minutes in the water but will be retained by the swimmer.

(c) Traverse 75 yards in deep water in full utility uniform (less cover) with boots/oxfords.

b. All water survival and swimming training will include instruction in the following:

(1) Employment of safety lines or other comparable expedients to assist in swift stream/river crossings.

(2) Adverse physiological effects caused by cold water (hypothermia) and the precautionary measures to be taken prior to exposure to such an environment.

(3) Employment of a standard and expedient flotation devices in a water survival situation.

(4) Removal of individual field equipment after unscheduled water entry. Entry will be made from a minimum height of 10 feet with a complete set of field equipment, properly packed as an expedient flotation device and waterproofed. The ALICE pack will then be removed, the rifle laid across the top of the pack, the helmet remaining or put back on the individual's head, and the individual will traverse 25 yards to the side of the pool or water's edge.

c. The additional water survival qualification (flight physiology training) and requalification required of personnel assigned to flight status will be conducted in accordance with OPNAVINST 3710.7_,Promulgation of NATOPS General Flight and Operating Instructions. Swimming qualification (first-class, SI) is a prerequisite to water survival (flight physiology) training and is a one-time requirement.

Section II. Water Sports and Competitive Activities

4201.GENERAL

The commander, when developing the combat readiness program, needs to look carefully at the mission. If the mission requires Marines who must perform mission-essential water-related activities of long duration, he may want to look to water sports/competitive activities as a method of developing endurance. Endurance and confidence can only be developed by extended time in the water. While defined training can accomplish this goal, the goal can be enhanced through training which is integrated with the physical conditioning program. This approach will provide variety and enthusiasm. Two activities which will accomplish this goal are water relays and water polo. The important consideration is that the unit achieve total participation by all members of the unit.

4202.WATER RELAYS

Development of water relays will generally conform to the guidelines in chapter 5, section IV. When contemplating relays, the commander should review chapter 5 and adapt the principles to the size of the unit and the facility to be used.

4203.WATER POLO

The official rules provided in this paragraph will need to be adapted to the unit's unique situation, its size, and the size of the facility. Commanders should make every effort to have all members of the unit participate.

a. NCAA Rules This style of play is based on two popular sports: swimming and basketball. Playing is done on the surface of the water by teams of seven players each. The size of the playing area is 25 yards by 14 yards, which is the size of the typical high school or college indoor pool. The deeper the water, the better. If the entire playing area is deep-6 feet or more—this is ideal, although most indoor pools have a shallow end. At each end of the pool is a goal. In deep water, the goal is 10 feet across and 3 feet high. In shallow water, where the goalie can stand on and jump from the bottom, the goal is 10 feet across and 5 feet high. One of the seven players on each team is the goalie. He is the only player on the team who can stand on or jump from the bottom (if it is shallow enough to permit this), catch and pass the ball with both hands at the same time, or hit the ball with a clenched fist. However, the goalie may not swim across or pass the ball across the middle of the pool.

b.Equipment. The goalie on the visiting team must wear a white cap with the number 1 clearly marked on it; his teammates, who are designated as guards and forwards, must wear white caps numbered 2 through 7, with the substitutes wearing white caps numbered 8 through 21.The goalie on the home team must wear a dark-colored cap with the number 22 clearly marked on it; his teammates designated as guards and forwards, must wear similar dark-colored caps numbered 29 through 42.The guards and forwards on each team may swim freely up and down the pool as they see fit, interchanging positions as often as they wish. They may not stand on or jump from the bottom, touch the ball with more than one hand at a time when catching or passing or shooting it, or enter inside the opposing team's 2-yard line unless preceded by the ball. The ball is similar to a soccer ball, except that it is yellow and covered with a waterproofing substance that makes it easier to handle with one hand.

c. Playing Time. In intercollegiate competition the game consists of four 7-minule quarters, the teams changing ends after every period of play. In high school competition the game consists of four-,five-, or six-minute quarters, depending on the ages of the participants. Here again the teams change ends after every period. There is a very brief interval between each quarter. At the start of each period the teams line up at their respective ends, and with ablast of the whistle, the head referee throws the ball into the middle of the pool. The fastest swimmers on each team then race to gain possession.

d. Officials. The head referee walks along one side of the pool watching for infractions. He is aided by the assistant referee, who patrols the opposite side of the pool. The referees are both equipped with whistles and two official's flags. These flags should be 12 inches square and should correspond in color to the caps worn by the respective teams. Whenever a referee sees an infraction, he blows his whistle and signals with his flag indicating that the player nearest to the ball with the appropriate colored flag will be given possession of the ball. Whenever the whistle is blown, all players should react by looking immediately to see which color flag is being waived which will indicate which team should be moving into the defense. Other officials are a timer and a scorer. As in all sports, it is important that the officials be skilled and competent.

e. Technical Fouls. The following are some of the common technical fouls: starting before the referee blows his whistle to open play; holding onto or pushing off from the side of the pool during play; taking hold of the ball underwater when tackled by an opposing player; swimming inside an opponent's 2-yard line unless preceded by the ball; touching the ball with both hands at the same time (goalie excepted); standing, walking on or jumping from the bottom when taking an active part in the game. When a technical foul occurs, thereferee blows the whistle and with his flag awards possession. The player on the team awarded possession nearest the point of infraction then has 5 seconds to put the ball into play by passing to a teammate or by dropping the ball into the water and swimming after it.

f. Personal Fouls. The following are some of the more common personal fouls: committing any of the technical fouls mentioned above for the purpose of scoring or preventing a goal; holding, ducking, pulling, pushing off from, swimming over, or impeding the arm or leg movement of any opponent who is not touching the ball; and splashing water in the face of an opponent. When a personal foul occurs, the referee will follow the same procedure as with a technical foul. While doing this, the referee will also call out clearly and loudly the number of the player who was guilty of the infraction and a personal foul will be marked against that player by the official scorer. When a player accumulates five personal fouls, he fouls out of the game and must be replaced by a substitute.

g. Penalty Shot. A penalty shot can be awarded by either of the officials or the scorer when:

o An offensive player inside the opponents' 4-yard line not touching the ball is held, ducked, pulled-back, kicked, or struck.
o A team has accumulated a total of ten personal fouls. When the former occurs, the head or assistant referee should immediately blow the whistle and by holding a two-flagged stick in a vertical position above his head, signal that a penalty shot has been awarded. When the latter occurs, the scorer should use a buzzer located at the scorer's tabic to signal that one team has accumulated ten personal fouls, thereby entitling the other team to a penalty shot. A penalty shot is taken from the 4-yard line in front of the goal. All players except the defending goalie must leave the 4-yard line until the shot is taken. No player can be within 1 yard of the shooter. After ascertaining that the shooting player is on the 4-yard line and the goalie is onThegoal line, the referee will ask the shooter to lift up the ball. When he docs, the referee will give a sharp blast from the whistle. At the whistle, the shooter must shoot without delay and without any faking at the goal. The goalie may try to block the shot. If he does, or the shot is otherwise missed, it is immediately in play and action continues.

h. When a Goal is Scored. When a goal is scored, either from a shot taken by a player out in the field or by penalty shot, the defending goalie must then pass the ball to the nearest referee. The referee will then pass the ball back to the goalie, who puts it into play by a pass to one of his teammates as soon as possible.

Section III. The Battle Swimming Test

4301.GENERAL

The battle fitness test is the evaluation contained in the Guidance for Basic Warrior Training (BWT) as part of the Basic Warrior Training Concept Plan.

4302.THE BATTLE SWIMMING TEST

The battle swim test consist of the following:
a. Be able to climb a three meter high board, while wearing a field uniform, carrying a M-16, web gear, and four canteens. Jump off blindfolded, recover and swim 100 meters with the M-16.
b. Repeat the same process, without the blindfold, and upon entering the water a second time, recover and retrieve all submerged gear.
c. Remain afloat and tread water for five minutes.
d. Demonstrate a knowledge of how to cross a stream using field expedient techniques and a single strand rope bridge.
e. Successfully complete helicopter egress training.

Chapter 5
COMPETITIVE CONDITIONING ACTIVITIES
Section I. Organization of Competitive Activities

5101.LEADERSHIP OF COMPETITIVE ACTIVITIES

a. Place in the Program. Competitive conditioning activities consist of dual combatives, relays, team contests, and team sports in which individuals or teams compete against an opponent to win. Competition is one of the best ways of maintaining interest in the physical training program during the sustaining stage. Organized competition provides enjoyable, vigorous physical activity that has proved to be one of the best supplements to conditioning drill activities. The benefits of competition are the development of aggressiveness, unit pride and identity, teamwork, and the will to win. These activities help develop cardiovascular and muscular endurance, strength, and coordination. Scheduling competitive activities in an orderly and progressive manner is desirable. The progression should be from relays to dual combatives, to team contests, and finally to team athletics. These activities should be conducted as part of the program after a basic level of conditioning has been developed, usually during the slow-improvement stage of conditioning. Muscles and joints should be strengthened by preconditioning to withstand the strain placed upon them by sudden stops and turns, body contact, bearing of weight, and falls. Competitive activities, however, should not be allowed to dominate the physical training program.

b. Area and Equipment. Some of the competitive activities included in this chapter require specific types of areas and equipment; others do not. The area requirement can usually be satisfied on available training fields. When items of equipment are required or specific courts or field layouts are to be marked off, such information is included.

c. Leadership. The principal factor for success as a leader of competitive activity is an energetic, dynamic, enthusiastic approach. The leader's attitude is reflected by the group, so the commander must carry on the activity in a snappy and vigorous manner. Confidence on the part of the commander will create an impression of decisiveness and certainty. Confidence grows out of experience and a thorough knowledge of the activity. Mastery of subject is the first step in developing confidence, assurance, and poise.

(1) The following suggestions are offered for leaders of competitive combatives, contests, and sports:

(a) Get the activity underway quickly by selecting and teaching only the essentials.

(b) Use rules to add to the enjoyment of the activity and do not allow them to interfere with the spirit of competition.

(c) Stop the activity before interest begins to lag.

(d) In team contests, clearly distinguish sides by attempting to maintain unit identity.

(e) Always insist on fair play, enforcing the rules impartially.

(2) The following procedure is recommended for presenting a competitive activity.

(a) Name the activity.

(b) Briefly explain the objective of the activity and give only the pertinent rules.
(c) Have a demonstration at slow speed and answer questions.
(d) Organize groups into teams and appoint captains.
(e) Arrange teams in the proper starting positions.
(f) Conduct the activity.

d. Competitive Units. Units for competition should be the same organizations in which Marines train: battalions, companies, platoons. In most situations, the unit is the squad.

e. Provisions for Instruction. One of the most effective methods of maintaining interest and participation in competitive activities is to provide instruction in activities with which most Marines are unacquainted. Such instruction can be conducted during regular physical training periods. Careful planning is required to keep all individuals continuously engaged in vigorous activity. The materials in the following sections should be used as a guide for instructional purposes.

f. Officiating and Control. Every effort should be made to provide good officials for all competitive activities. Poor officials quickly cause dissatisfaction among participants in team activities and create a situation which can turn an organized activity into a brawl. Each company should have or develop several competent and qualified officials available for games on company and platoon levels.

5102.MILITARY FIELD MEETS

When units reach the latter part of the slow-improvement and sustaining stage of training, interest in the program may lag. A change in course content can arouse the desire to participate. An event which does not require a high degree of skill yet demands strenuous activity is ideal for the military field meet. This is a series of team contests conducted on a station-to-station basis during a given period of time. Team contests carried on simultaneously provide essential training, vigorous exercise, and stimulating competition in an atmosphere that is enjoyable for all. Because of the healthy rivalry that a field meet arouses, it is an excellent form of inter-unit competition.

a. Objective and Advantages. The objective of the military field meet is to provide activity for everyone in the participating units. Activities are chosen that will develop aggressiveness, teamwork, a will to win, and competitive spirit, and that will stimulate interest and build esprit de corps. The military field meet can be included as part of the physical training program or as part of the off-duty recreational program. It is a form of contest that can be conducted in nearly all circumstances because it can be easily modified, requires a minimum amount of equipment, and can be readily organized.

b. Level of Competition. The military field meet is flexible. It can be adjusted to large or small groups. A company-size unit is the most desirable, but it may also be administered within a larger unit. If it is conducted within a company, the participating units will be squads. If it is held within a larger unit, platoons compose the teams. A larger-size unit requires more extensive organization and administration.

c. Selection of Events. In organizing a military field meet, select events that are simple and easy to administer. All rules and regulations should be clearly understood by everyone. No event should require previous practice. In selecting the events, consider the interest and capabilities of the Marines and available equipment and facilities. Select events which will require that all members of the units participate. Events should not be dominated by the athletes; rather, MOS-related events which have significance in producing skills and combat readiness should be used.

d.Equipment and Facilities. The site must be large enough to permit events to be grouped about a central control point. Use available facilities such as parade decks, Softball fields, or tracks. A public address system is desirable at the control point for the initial orientation of teams and for subsequent announcements of time lapses, cumulative scores, and final standings. A tally board is necessary so that all units can constantly monitor how the events are progressing.

e.Administrators. Efficient administration of the military field meet depends on the referees, judges, and scorers. For these positions, choose individuals within the units who have had athletic or officiating experience. Prior to the day of the event, all administrators should be briefed and assigned a specific task in order that they may become familiar with the rules and organization of the contests they will conduct. The following officials should be available:

(1) A primary instructor or supervisor who is in charge of the control center and who is responsible for the successful operation of the athletic carnival. The supervisor must have an assistant to act as a timer and scorer.

(2) One assistant instructor in charge of each event. This person should be-

(a) Familiar with the rules of the event.

(b) Effective as a leader to ensure proper supervision and control over the participating teams.

(c) Enthusiastic to provide proper motivation.

(d) Self-confident of the ability to judge infractions of the rules. The assistant instructor must be fair in judgment and penalize without hesitation when infractions occur.

(3) Runners between event stations and the control point. These individuals collect and deliver scores.

f. Team Organization for Competition. The size of the teams is determined by the level on which the field meet is organized. Maintaining the integrity of the unit promotes esprit de corps, but this does not preclude grouping two squads into one team.

g.Conduct of Events

(1) The assistant instructor at each station takes charge of the group and gives a brief explanation of the major rules of the event. The assistant instructor speaks clearly and distinctly from a position to be seen and heard by all. A short demonstration is desirable if it will help clarify the event.

(2) The assistant instructor should make certain that teams can be clearly distinguished by the use of, for example, T-shirts and fatigue jackets, or caps and no cap, or colored jerseys. The assistant instructor should teach a whistle response (teams stop play immediately upon hearing whistle), get the event started as quickly as possible, and make any necessary corrections as the contest progresses. The rale of good officiating is to use a minimum of calls, yet maintain control of the contest. Penalize when necessary, but refrain from disqualifying contestants or teams.

(3) Keep the activity moving as rapidly as possible. When the central control point sounds the whistle to stop the play, all competition ends immediately. The assistant instructor then assembles the group, forwards the team scores to the central point, and, upon a signal from the central control point, rotates the teams to their next station. It is essential that rotation and orientation of teams be quick and orderly to reduce time spent between contests.

(4) Upon completion of the final event, assistant instructors move their teams to the central control point for the announcement of winning teams, presentation of awards (if any), and final critique.

h. Scoring System. The system for determining the winner of the military field meet should be simple and efficient. At the completion of competition, the scorer totals the points that each team has scored in all contests. The scorer then subtracts the number of points scored against a team (penalty points for nonparticipation,etc.) from the total. The resulting scores are placed in a column with the highest score at the top and the lowest at the bottom. The team with the highest total is the winner. This type of scoring system encourages a team to ensure total unit participation.

Section II.Combatives

5201.DESCRIPTTON AND OBJECTIVE

Combatives are strenuous, short competitive contests in which two individuals attempt to overcome each other in a bout of skill and strength. These contests help to develop Marines' resourcefulness, confidence, strength, agility, coordination, and will to win. Any level ground area can be used. However, extremely hard ground should be avoided as some of the combatives require ground contact. A whistle is needed to control the bouts during the competition since voice commands may go unheeded. The extended rectangular formation is used for dual combatives. Combatives are conducted on an informal basis. Marines are allowed to remain at ease between activities and are allowed to brush themselves off after being on the ground. To get the most out of combatives, individuals must be urged to overcome their opponents as quickly as possible as would be required in combat.

a. Benefits for Marine Training. The possibility of close contact with an enemy in combat faces Marines at all times. They must be trained to react aggressively and violently. Combatives may be used as an introduction to such hand-to-hand contact and should be followed by hand-to-hand combat training. Marines enjoy competition, and this type of activity is a welcome change from formal conditioning activities.

b. Instructor Responsibility

(1) The instructor tells the individuals that all combatives begin and end on the whistle signal. The instructor demonstrates each activity before having it performed, explaining it in simple terms.

(2) After stopping one activity, the instructor gets everyone in place for the next.

(3) The instructor must closely supervise combatives to insure that contestants do not use unfair or unsportsmanlike tactics. To avoid unnecessary injury, instructors must see that the bouts are closely controlled and opponents equally paired. Adjustments should be made in apparent cases of mismatched abilities.

5202.COMBATIVES TABLES

There are three tables of combatives, each of which can be completed in 15 minutes. These tables become progressively more difficult from lower to higher numbers.

a.Combatives Table I

(1) Open-Hand Slap Boxing(See fig. 5-1, A. individuals assume a boxer's stance, palms open, fingers extended and joined. Each contestant tries to slap the opponent about the head and upper body with the open hand. This is a good warm- up activity.

(2) Wrist Tug-of-War(See fig. 5-1, B.)Two individuals sit on the ground with the soles of their feet in contact. Each grasps the opponent's wrists so that the hands are directly over the feet. At the whistle, each individual tries to pull the opponent from sitting to standing position.

(3) Arm-Lock Wrestling.(See fig. 5-1, C.)Two individuals sit back-to-back with legs spread and arms locked at the elbows. Each right arm is inside the opponent's left. Each tries to force the opponent's left arm or shoulder to the ground. The individual who first wins three bouts is the winner.

(4) Bulling(See fig. 5-1, D.)Two individuals assume the Westmoreland wrestling hold, each grasping the opponent's neck with the right hand and the opponent's right elbow with the left hand. Each tries to force the other to move one foot by pushing, pulling, or otherwise manipulating. The individual who first wins two boutsis the winner.

(5) Indian Wrestling(See fig. 5-1, E.)Two individuals lie on the ground, side by side, with their heads in opposite directions. They link right elbows. On the instructor's signal or by mutual agreement, each individual raises the right leg approximately straight and far enough to engage the opponent's heel. To start the contest, each person usually raises the leg three times rhythmically and on the third time engages the opponent's heel. Each tries to roll the other over backward. The right leg is used for three bouts, then the left leg for three bouts.

Figure 5-1 is available in the illustration section at the end of the book.

b. Combatives Table II

(1) Wrist Bending.(See fig. 5-2, A.) Opponents pair off, face each other, raise their arms forward, and, with palms forward, interlock their fingers. At the starting signal, each individual attempts to bend the opponent's wrist. The arms are kept up and forward and are not allowed to swing around and down to the sides. The individual who first wins two bouts is the winner.

(2) Back-to-Back Push.(See fig. 5-2, B.)Two individuals stand back-to-back with elbows locked. Each right arm is inside the opponent's left arm. At the starting signal, each pushes backward, trying to move the opponent. Opponents are not allowed to lift and carry each other; only pushing is permitted. The one who pushes the opponent the farthest wins the bout. The individual who first wins two bouts is the winner.

(3) Hop and Pull Hand.(See fig. 5-2, C.) The individuals are matched in pairs. Each grasps the opponent's right hand and, hopping on the right foot, attempts to pull the opponent off balance. Contestants automatically lose if they touch their free hand or their lifted foot to the ground. For successive bouts, they alternate hands and feet.

(4) Westmoreland Wrestling. (See fig. 5-2, D.)Each contestant grasps the back of the opponent's neck with the right hand and the opponent's right elbow with the left hand. In this position, each attempts to pull, push, or force the opponent to touch the ground with any part of the body other than the feet. The individual who first wins two bouts is the winner.

(5) Crab Fight.(See fig. 5-2, E.)Two individuals sit on the ground facing in opposite directions with their hands on the ground behind them. At the whistle, they raise their hips and push with their shoulders and bodies, each trying to make the other's hips touch the ground. The individual who first wins two bouts is the winner.

Figure 5-2 is available in the illustration section at the end of the book.

c. Combatives Table III

(1) Hand Wrestling.(See fig. 5-3, A.) Two individuals stand facing each other. Their right feet are forward and braced side by side. They grasp right hands for the first bout, left hands for the second bout. Each pulls, pushes, makes sideward movements, and otherwise maneuvers to force the opponent to move one or both feet from the original position. The contestant who first wins two bouts is the winner.

(2) Back-to-Back Tug.(See fig, 5-3, B.) Two individuals stand back-to-back with both arms linked at the elbows. Each has the right arm inside the opponent's left arm. At the starting signal, each attempts to pull the opponent. Lifting and carrying are permitted. The contestants must maintain their original direction and keep their arms linked. After a predetermined time, the player pulled or carried the farthest is the loser.

(3) Wrestling to Lift off Feet.(See fig. 5-3, C.) Contestants face each other. Each places the right arm under the left arm of the opponent and around the body. The left arm is over the opponent's right shoulder. Each tries to lift the other off the ground. The individual who first wins two bouts is the winner.

(4) Arm Pull Between Legs.(See fig. 5-3, D.) Two individuals are paired off, back-to-back. Each bends forward and, extending the right arm between the legs, grasps the opponent's right wrist. At the starting signal, each person attempts to pull the opponent After a predetermined time, the player who has pulled the opponent the farthest is the winner of the bout. The one who first wins two bouts is the winner. Repeat with the left hand and then both hands.

(5) Rooster Fight.(See fig.5-3, E.) Each contestant grasps the left foot with the right hand from behind, and the right arm with left hand. Each hops on the right foot, and, by shoulder-butting the opponent or by feinting and sudden evasions, forces the opponent to let go of the foot or arm. The contestant who first wins two bouts is winner.

Figure 5-3 is available in the illustration section at the end of the book.

Section III .Relays

5301. DESCRIPTION AND OBJECTIVE

Relays are races in which each member of a team runs one leg of the race. The team effort decides the winner. Relays provide stimulating competition and contribute to the conditioning of personnel. They also develop aggressiveness, team spirit, and the will to win. Relays should be dispersed throughout the program for short periods of time to provide a change of activity.

a. Team Organization. Relays are conducted most efficiently in platoon-size groups. Teams of equal size must be organized. Competitive spirit is encouraged and teams are organized better by basing teams on units such as squads, crews, or sections. Team captains should be designated. Extra Marines may be used as officials. The number of individuals on a relay team should be limited to squad size. If larger teams are used, the runners will spend too much time awaiting their turns and too little time actually participating. Two to six teams are ideal for relay competition. It is difficult to keep track of winners when more teams compete.

b. Administration of Relays

(1) The time spent on any one relay should be relatively short. If one team achieves a substantial lead in a long relay, the competitive spirit and enthusiastic participation of the other teams may decrease. Several short relays are generally better than one long relay.

(2) To maintain competitive spirit throughout a number of relays, determine the teams that win, place, and show in each relay and their total points for all relays. This can be done by awarding points to all teams on the basis of position at the finish of each relay. The team with the greatest number of points is the winner of the entire set of relays.

(3) Difficulties commonly encountered in conducting relays may be avoided by the following procedures:

(a) The last player in a relay race should be conspicuously identified. For example, the last runner can tie a handkerchief around the head or arm, take off the shirt, put on a hat or take it off, or use some other means.

(b) Another way to keep track of the progress of the race is to have each player sit or squat as soon as each is finished.

(c) Judges at the starting line can keep the runners from starting too soon.

(d) To prevent contestants from turning before they run the full distance, they should be required to run around a peg, pole, or assistant instructor.

(e) Batons, handkerchiefs, tent pegs, or other objects should be passed from one runner to the next when relays are run on a circular track.

(4) Before a unit's first participation in a relay, inform participants of the rules and scoring system. Violation of the rules should not result in disqualification. Instead, impose point penalties. A point penalty is imposed by subtracting a point from the team total at the conclusion of the relay.

(5) Careful administration will prevent most violations. For efficient conduct of relays, follow this procedure:

(a) Announce the name of the relay.

(b) Form the individuals in relay position.

(c) Briefly explain the relay and the rules for running it.

(d) Demonstrate.

(e) Have a definite finish line, and ensure that all know where it is.

(f) Answer questions.

(g) Conduct the relay.

(h) Determine winner and award points.

5302. RELAY TABLES

a. Events. The following relays are grouped into a table of activities. Each relay table can be completed in 15 minutes. Thus, relays can be used as a sole activity or as a part of a longer period. Each table is planned for a platoon-size group (30-60Marines). Adequate warmup for participants is provided by conducting several repetitions of exercises 1 and 2 of a conditioning drill. The recommended relay tables require an area 40 by 60 meters in size. Each table provides a variety of activity. The tables are progressive in the overload applied and should be scheduled in numerical order although not necessarily on successive days.

b. Relay Table I

(1) **60-Meter Lane Relay** (See fig. 5-4, A.) Each team is assembled in single file behind the starting line. On signal, the first individual of each team runs to the turn-around line 30 meters away, then runs back and touches the next teammate waiting at the starting line. The winning team is the first team to get its last member across the finish line. If an individual starts before being touched by the preceding runner, the team may be penalized.

(2) Wheelbarrow Race.(See fig. 5-4, B.) The players of each team pair off and line up in single file behind the starting line. The first individual walks on the hands while the partner grasps the ankles. They advance to the distance line (25 meters) behind which they exchange positions and return to the starting line. The rear individual must always hold the partner's ankles. After the first pair returns across the starting line, the next pair starts.

Figure 5-4 is available in the illustration section at the end of the book.

(3) Squad Front Relay .(See fig. 5-4, C.) The teams form in a line along the starting line with a 10-foot interval between teams. The members of each team lock elbows so that they are linked together. At the starting signal, the teams run to the distance line (20meters) where the left flank individual acts as a pivot. The team swings around on the pivot and returns to the base line. If a team breaks its links, it must reform before continuing. The first team to completely cross the base line intact is the winner.

(4) Crab-Walk Race.(See fig. 5-4, D.) The players of each team line up in single file. The first person of each team assumes the crab-walk position with feet forward on the starting line, hands on the ground behind it. At the starting signal, contestants move forward to the distance line (10 meters). They touch the line with their feet and then return to the starting line in the reverse position with the head and hands leading. The second person may not start until the first individual touches the finish line.

c.Relay Table II

(1) 100-Meter Lane Relay.(See fig. 5-5, A.)This relay is conducted exactly as the first relay of table 1 except that the start and turn-around line are 50 meters apart. This relay provides progression in sprinting.

(2) Frog-Jump Relay. (See fig. 5-5, B.) Each team lines up in single file. The first individual assumes a squatting position on the starting line. At the starting signal, the contestant progresses to the distance line (15 meters) and back by leaping forward, catching the weight on the hands, and bringing up the legs to the squat position for the next leap forward.

(3) Simple Relay. (See fig. 5-5, C.)Each team lines up in single file. Place a marker on the distance line (20 meters) in front of each team. Each team member, behind the first one, grasps the belt of the individual ahead. At the starting signal, each team runs as a unit to the marker, circles it, and returns to the starting line. The first team to completely cross the starting line intact is thewinner.

(4) Fireman's Carry Relay.(See fig. 5-5, D.)The players of each team line up in pairs, one behind the other. One individual in each pair carries the partner to the distance line (30 meters), using the fireman's carry. At the distance line, individuals exchange places and return to the starting line. As a variation, the person to be carried lies on the ground. The one carrying lifts the partner to the proper position. This relay may be performed with the other carries described in guerrilla exercises in chapter 3.

d.Relay Table III

(1) 200-Meter Circle Relay .(See fig. 5-6, A.)A course is laid out in a circular, rectangular,or oval pattern that is 200 meters around. Each team provides one runner on the starting line. On signal, the runner races around the 200-meter track and touches the next teammate, waiting at the starting line, who runs the same course. Each team member runs one lap of the course.

Figure 5-5 is available in the illustration section at the end of the book.

(2) Bear-and-Crab Race.(See fig. 5-6, B.) Each team lines up in single file. At the signal to start, the first contestant in each column assumes the bear-walk position (hands on ground in front), walks to the distance line (15 meters), and then runs back to the starting line. The returning contestant touches off the second individual and goes to the rear of the line. The second contestant assumes the crab-walk position (hands on ground behind) and crab-walks with feet leading to the distance line. This contestant also runs back to the starting line and touches off the next individual, who walks bear fashion. The rest of the members of each team alternate in this manner. The relay ends when the first person is back at the head of the line.

(3) Pilot Relay.(See fig. 5-6, C.) The players are grouped in threes, arms interlocked at the elbows, and outside players facing backwards. The middle person runs forward; the two outside individuals run backward. They run to the turning point (15meters), where they start back, this time with the middle person running backward and the two outside individuals running forward. The next set of three players starts when the first set crosses the starting line.

(4) Saddle-Back Relay.(See fig. 5-6, D.) Mark two parallel lines 15 meters apart. Each team selects a rider. The remaining members of each team count off. The even-numbered players from each team form in single files behind one line and the odd-numbered players from each team form in single files behind the other line directly across from their teammates. At the starting signal, the rider mounts the back of the first player of the team who carries the rider across the other line. There the rider changes mounts to the first player in the second line without touching the ground. This person carries the rider to the next player waiting in the first line. The relay continues until all of the mounts have carried the rider. A rider who falls off must mount again at the point of the fall. A rider who falls in changing mounts must get back on the original mount before making the change.

e. Relay Table IV

(1) 100-Meter Circle Relay (See fig. 5-7, A.)A course is laid out in either a circular, rectangular, or oval pattern that is 200 meters around. Each team is divided in half with each half positioned at starting lines on opposite sides of the track. Each runner races halfway around the track and touches a teammate who completes the lap. Each runner then waits in file at the halfway point until touched by the runner and then completes the second half of the lap. The first team to return all runners to their original starting line is the winner.

Figure 5-6 is available in the illustration section at the end of the book.

(2) In-and-Out Relay.(See fig. 5-7, B.) Each team lines up in a file with players 2 meters apart. At the starting signal, the first player runs back through the column in a zigzag fashion. He alternates, going to the right of one teammate and to the left of the next. Upon completing the run, the player lines up 2 meters behind the last one in the column. As soon as the first runner has passed the second person, the latter starts to run. This continues until all the players have realigned their original order. The team that finishes first is the winner. It may be desirable to have this relay continue until all individuals have run through their entire team two or three times in succession.

(3) Circle Race.(See fig. 5-7, **C.**) Each team forms a circle and holds hands with all individuals facing out except one who faces in and is the "driver."At the starting signal, the teams race to the distance line (20 meters) and back, keeping the circle intact. All the individuals in the circle must completely cross the distance line. The "driver" gives directions and orders. When the circle breaks, it must be reformed before it can continue. The first team completely over the starting line is the winner.

(4) Horse-and-Rider Relay.(See fig. 5-7, D.)Each team lines up in a single file. At the signal to start, the second player in each column leaps upon the back of the first who carries the rider across the distance line (30 meters). At the distance line, the rider dismounts and runs back to the starting line. There the rider picks up the third individual in the column, and carries this player to the distance line where the first player has remained. This continues until the last person is carried across the distance line.

Figure 5-7 is available in the illustration section at the end of the book.

Section IV. Team Contests and Athletics

5401. DESCRIPTION AND OBJECTIVE

Team contests are competitive activities in which Marines as a team compete with another team to win. They are guided by simple rules and organization. The function of team contests is to provide competition and an opportunity for body contact and to contribute to the development of physical readiness. In competing and working together as a team, individuals develop aggressiveness, the will to win, and teamwork. Team athletics deserve a prominent place in the physical training program because they contribute to increased combat efficiency. Because of the competitive nature of athletics and their natural appeal, individuals take part with enthusiasm. Athletic teams formed at intramural and higher levels are a strong unifying influence and provide one of the best means of developing esprit de corps.

a. Area and Equipment A level training field is sometimes the only area required. Many contests need no equipment. In contests requiring equipment, the need is for standard items such as logs, balls, nets, goals, and similar types of equipment. Specific requirements for area and equipment are listed with each contest description.

b. Progression. Team contests of a strenuous nature should be introduced after a basic period of conditioning has been completed and individuals are in the slow improvement stage of conditioning. Progression can take place from the less active to the more vigorous contests, and then from the non-contact to the contact or combative-type contests.

c. Necessity for Preconditioning Individuals must undergo conditioning prior to participation in athletics. Muscles, organs, joints, and ligaments not accustomed to stress and strain from sudden stops and starts, falls, body contact, rapid turns, prolonged running, and other rigors of athletic competition are subject to injury. Although athletics should not be introduced until players are physically prepared, there is still opportunity to engage in competition through lead-up contests. Individuals learn many of the skills required for athletics while participating in team contests.

d. Benefits. Athletics are beneficial primarily to sustain interest in the program and to maintain an achieved level of physical fitness. Athletics are a supplement, not a substitute for other types of conditioning activities which should continue. All of the desirable traits of physical fitness cannot be developed through athletics, yet their contribution is significant. For athletics to make a proper contribution to physical conditioning, the selected sports must bevigorous. The team contests recommended here are carefully selected for their simplicity, aggressiveness and applicability to squad- teaching teamwork. They are generally strenuous and involve the whole team throughout the entire duration of the contest.

5402. TEAM CONTESTS

a. Pushball

(1) **Players.** 10 to 50 players on a side. (See fig. 5-8.)

(2) Equipment A large pushball 5 to 6 feet in diameter.

(3) Area. A field 240 to 300 feet long, 120 to 150 feet wide. Mark a center line in the middle of the field parallel to the end lines. Mark a line 45 feet on either side of this center line and parallel to it, extending it across the width of the field. Mark another parallel line 15 feet from each end line, extending it across the width of the field.

(4) Game. Four 10-minute quarters are played. Give 2-minute rests between quarters and 5-minute rests between halves. The object of the game is to propel the ball over the opponent's goal line by pushing, rolling, passing, carrying, or any other way except kicking.

(a) The ball is placed on the center line with the opposing captains 3 feet from the ball. The rest of the players are 45 feet from the ball, on their half of the field. On the referee's starting whistle, the captains immediately play the ball with their respective teams coming to their assistance.

(b) At quarter time, the ball remains dead for 2 minutes at the spot where it was when the quarter ended. At half time, the teams exchange goals. The play is then started as it was in the beginning.

Figure 5-8 is available in the illustration section at the end of the book.

(c) Players may use any means of interfering with an opponent's progress except striking and clipping. Clipping is throwing the body across the back of an opponent's leg, while the opponent is running or standing. Legal use of force may be applied to all opponents whether or not they are playing the ball. For striking an opponent, the offender is removed from the game. The team penalty is half the distance to their goal. The penalty for clipping is the same.

(d) When any part of the ball goes out of bounds, it is a dead ball. The teams line up at right angles to the side lines and 3 feet apart at the point where the ball went out. The referee then tosses the ball between the teams.

(e) When, for any reason, the ball remains in one spot for more than 10 seconds, the referee declares the ball dead. The ball is then put into play as it is for an out-of-bounds situation.

(5) Scoring. A goal is scored when the ball, or any part of it, is propelled across the opponent's end line. A goal counts 5 points. The team scoring a goal has the privilege of trying for a point after the goal. To try for this extra point, the ball is placed on the opponent's 5-yard (or 15-foot) line. The teams line up on either side of this line separated by the width of the ball. Before the whistle blows, one player on each team may place hands on the ball. On the referee's signal, the ball is put into play for 1 minute. If any part of the ball is driven across the goal line in this 1-minute period, the offensive team scores 1 point. The defending team may not score during the opponent's try for the extra point.

b. Line Rush

(1) Players. Any number up to 50 on each side. (See fig. 5-9.)

(2) Equipment. None.

(3) Area. A field, 75 by 100 feet.

(4) Game. One team lines up behind one goal line and the other in midfield. On the starting signal, the team standing behind the goal line seeks to cross the goal line at the opposite end of the field within 30 seconds. The team in the center seeks to prevent it by catching and holding the runners. At the end of 30 seconds, the teams change.

(5) Scoring. Count the number of individuals who have crossed the far goal at the end of 30 seconds. After each team has had from three to five tries, the scores are added and the winner declared. A player scores 1 point when any part of the body is across the goal line.

Figure 5-9 is available in the illustration section at the end of the book.

c. Human Tug-of-War

(1) Players. 10 to 20 on a team.(See fig. 5-10.)

(2) Equipment. None.

(3) Area.40 to 60 feet.

(4) Game. Draw a line in the center of the area. Divide the players into two equal teams. Place them in single file on opposite sides of the center line facing each other. Each individual places the arms around the waist of the teammate in front. The two leaders of the opposing teams grasp each other around the waist. On signal, the teams try to pull each other over the center line within 30 seconds.

(5) Scoring. The team pulled across the center line loses. If neither team is pulled over the center line, but one team breaks its file, that team loses the match.

(6) Variation. Use a 3/4- or 1-inch rope. Space the leading players on each team 10 feet apart, each holding one end of the rope. The team pulled across the center line loses.

Figure 5-10 is available in the illustration section at the end of the book.

d. Master of the Ring

(1) Players. Any number. (See fig. 5-11.)

(2) Equipment. None.

(3) Area. A clearly marked circle large enough to contain all the players.

(4) Game. All the players stand inside the circle. At the signal, all players attempt to throw each other out of the circle. All tactics are fair except unnecessary roughness. When any part of the body touches across the line, the player is out and leaves the circle at once. Several officials are needed to spot the players who cross the line.

(5) Scoring The player who remains in the circle when all the others are out is the master of the ring.

(6) Variation 1. The players are divided into two equal teams. Each team is clearly marked. On signal, each team tries to throw the opponents out of the circle. The winning team is the team that eliminates all opponents from the circle.

(7) Variation 2.The players are divided into two equal teams. Each team sends only one individual into the circle. When one person has been forced out of the circle, the losing side sends in only one player. The team which eliminates all opponents is the winner.

(8) Variation 3.A pit, approximately 4-feet deep, is used rather than a circle on level ground. The contest may then be an individual or team activity.

Figure 5-11 is available in the illustration section at the end of the book.

e. Log Pivot Circle. Space teams far enough apart so each team can pivot in a circle without colliding with any other team. Each team holds a log in the bend of the arms in front of the chest. (See fig.5-12.) At the command **CIRCLE RIGHT, MOVE**,the left-flank individual holds the pivot while the log is carried around 360 degrees in a counter- clockwise motion, back to the original position. This movement may also be performed to the left by facing in the other direction and pivoting clockwise(**CIRCLE RIGHT, MOVE**).Other commands may be used such as: **CIRCLE HALF-RIGHT, CIRCLE HALF- LEFT** and so on. The first team to complete the prescribed movement is the winner.

Figure 5-12 is available in the illustration section at the end of the book.

f. Log-Rolling Race. Each team tries to roll its log a measured distance by pushing it with the hands and driving the body forward with the legs. (See fig. 5-13.)The first team to get the entire length of the log across the finish line wins.

Figure 5-13 is available in the illustration section at the end of the book.

g. Prone Pushing Contest.. Two teams lie prone, facing each other with a log between them. (See fig. 5-14.)Both teams place their hands against the log, keeping their arms straight. Then, by driving with the legs, each team attempts to push the other a measured distance to the rear.

Figure 5-14 is available in the illustration section at the end of the book.

h. Shuttle Relay Race. Each team in this race divides into two parts, A and B. Members of part A run 50 yards with a log held under their right arms. (See fig 5-15.)At the distance line, they give the log to their teammates in part B who bring it back to the starting line. The team pair finishing first is the winner.

Figure 5-15 is available in the illustration section at the end of the book.

5403.CROSS-COUNTRY AND DISTANCE RUNNING

a. Value of Running. Long-distance running gives some benefits that cannot be obtained in the same degree from any other sport. It builds powerful leg muscles, increases lung capacity, and develops endurance. For these reasons, cross-country and distance running should be included in the physical training program. These sports require only a few miles of open space which is generally available. Short cross-country runs and middle-distance runs can be used to supplement other activities, particularly team sports or the sports that develop precision or agility rather than endurance. Short cross-country runs can be scheduled once a week, gradually increasing the distance as the physical condition of the runners improves.

b. Cross-Country Runs. A cross-country run is a distance run held on a course laid out along roads, across fields, over hills, through woods, and on any irregular ground. A flat cinder or dirt track is not a suitable surface for cross-country running. Some runs are as long as 6 miles while others are as short as 3 or 4 miles. If cross-country running is being used to supplement other activities, the 3-mile course is long enough for most. Cross-country runs should be used only after individuals reach the sustaining stage of conditioning. This type of running should then be scheduled occasionally to provide variety. Cross-country running has the advantage of allowing mass participation. Interest can be stimulated by putting the runs on a competitive basis.

c. Practice Methods. Conditioning is essential to distance and cross-country running. Championship distance running depends on stamina, which can be developed only through constant training. An individual of only average ability can become an outstanding distance runner by steady and careful training. Hiking is an excellent method for getting into condition before the season opens. Long walks build up leg muscles. During the first month of the season, training should be gradual, starting with short distances, and increasing day by day. At first the legs will become stiff, but the stiffness gradually disappears if running is practiced for a while every day. To prevent strain, it is essential to limber up thoroughly each day before running.

(1) In training a large group, leaders should be stationed at the head and the rear of the column. They should make every effort to keep the runners together. After determining the abilities of the unit in cross-country running, it is advisable to divide them into three groups. The poorest-conditioned group is started first, the best-conditioned group, last. The starting time of the groups should be staggered so that all of them come in at about the same time.

(2) In preliminary training, running is similar to ordinary road work in that it begins with rather slow jogging, alternating with walking. The speed and distance of the run is gradually increased. As conditioning improves, occasional sprints may be introduced. Atfirst, the distance run is from one-half to 1 mile. It gradually increases to 2 or 3 miles. On completing runs, require runners to walk for 3 or 4 minutes before stopping to permit a gradual cooling off.

d. Facilities and Equipment. There should be at least one timer with a stopwatch (preferably three) for timing the runners. A course 3 or 5 miles long should be measured and marked by one of the three following methods:

(1) Fasten directional arrows to the top of tall posts and place them at every point where the course turns. Such signs should also be placed at every other point where there may be doubt as to the direction of travel.

(2) Place a lime line on the ground over the entire course.

(3) Place flags, clearly visible to the runners.

- **(a)** A red flag indicates a left turn.
- **(b)** A white flag indicates a right turn.
- **(c)** A blue flag indicates the course is straight ahead.

e. Rules/Scoring. A cross-country team consists of seven individuals, unless otherwise agreed. In dual meets, a maximum of 12 on one team may be entered, but 7 or fewer are scored. First place scores 1 point, second place 2, third place 3, and so on. All who finish the course are ranked and tallied in this manner. The team score is then determined by totaling the points scored by the first five of each team to finish. The team scoring the least number of points is the winner. If fewer than five (or the number determined prior to the race) finish, the places of all members of that team are disregarded. If two or more teams score the same number of points, the event is called a tie.

5404.TEAM ATHLETICS

Team athletics add variety to the combat readiness program. At no time should they be allowed to replace intramural activities or to provide practice time for intramural or base teams. If team athletics are integrated into the program, the goal should be total participation and competition. Therefore, the games selected can be played by the skilled or the unskilled. They require limited equipment, yet will teach competitiveness, aggressiveness, and teamwork while they improve conditioning.

a.Soccer. Soccer is one of the best athletic activities for developing endurance, agility, leg strength, and great skill fulness in using the legs. The most popular sport in the world, soccer is the national game of many European, Asian, and Central and South American countries. In recent years, it has become popular in the United States. A soccer ball is the only equipment needed for the game, and Marines can learn to play it easily. Players do not need much skill to participate, but the amount of skill they can develop is almost unlimited.

(1) Place in the Program. Soccer should be introduced during the latter part of the slow improvement stage of physical training. It can be used as a competitive activity in the sustaining stage. It is primarily a spring or fall sport. Any level field is suitable for competition. The boundaries for the soccer field are similar to the dimensions for a football field. (See fig. 5-16.) Goal posts are essential, but they are easily constructed and usually temporary so that they may be removed when not in use.

Figure 5-16 is available in the illustration section at the end of the book.

(2) Basic Skills

(a) Passing. Passing with the feet is the primary means of moving the ball. Short passes are easier to control. Emphasis should be continually placed on the skill of passing.

(b) Dribbling. The ball is dribbled by a series of kicks with the inside or outside of the foot. Do not kick with the toe. Keep the head over the ball when kicking and propel it only a short distance at a time. Keep it close to the feet. If a player lets a ball move far from the feet, an opposing player can easily take it away.

(c) Instep Kicking. The instep kick is the basic soccer kick. The toe does not come in contact with the ball. The toe is pointed downward, and the instep (the area under the shoe laces) is applied to the ball with a vigorous snap from the knee.

(d) Inside-of-the-Foot Kicking. The ball is kicked with the inside of the foot, and the leg is swung from the hip. The toe is turned outward, and the sole of the foot is parallel with the ground as the foot strikes the ball. This kick is used for short passes and for dribbling.

(e) Foot Trapping. The foot trap is the method of stopping the ball by trapping it between the ground and the foot. Place the sole of the foot on top of the ball at the instant it touches the ground, keeping the foot relaxed. This is an effective way to stop a fast-moving ball.

(f) Shin Trapping. .The shin trap is a method of stopping the ball with the shins. Stand just forward of the spot where the ball should strike the ground and allow it to strike the shins in flight or on the bounce. Use either one or both legs from the knee down so as to absorb the bounce.

(g) Body Trapping. .Intercept the ball with any part of the upper body except the arms and hands. Keep the body relaxed and inclined toward the ball. To keep the ball from bouncing, move backwards from it as it strikes the body. This drops the ball at the feet in position for dribbling or passing.

(h) Heading. Heading is a technique for changing the direction of the flight of a ball by butting it with the head. Tense the neck muscles and jump up to meet the ball. Butt the ball with the forehead at about the hairline to reverse its direction. Use the side of the head to deflect it to the side.

(3) Offensive and Defensive Positions. Forwards usually play on the offensive half of the field. Fullbacks usually play on the defensive half of the field. Halfbacks are the backbone of the team. They move forward on the offense and back on defense. The goalkeeper almost always remains within a few feet of the goal.

(4) Abridged Rules

(a) A soccer team is composed of 11 players. (See fig. 5-16.)

(b) The player propels the ball by kicking it with the feet or any part of the legs, by butting it with the head, and by hitting it with any portion of the body except the arms or hands.

(c) The goalkeeper is the only individual allowed to place hands on the ball, but only in the goalkeeper's area. The term "hands" includes the whole arm from the point of the shoulder down.

(d) A goal is made by causing the ball to cross completely the section of the goal line lying between the uprights and under the cross bar.

(e) Each goal scores 1 point for the team scoring the goal.

(f) The penalty for a foul committed anywhere on the playing field (except by the defensive team in its penalty area) is a free kick awarded to the opposing team.

(g) All opponents must be at least 10 yards from the ball when a free kick is taken.

(h) The penalty for a foul committed by the defensive team in its penalty area is a penalty kick.

(i) A penalty kick is a free kick at the goal from the spot 12 yards directly in front of the goal. The only players allowed within the penalty area at the time of the kick are the kicker and the defending goalkeeper.

(j) An official game consists of two 30-minute halves,

(k) Teams change goals at the end of every quarter.

(l) In the event of a tie, an extra quarter may played. This may be followed by a sudden death period. If the tie still exists, the game can be decided by penalty kick.

(m) After a team has propelled the ball across a side line and out of play, the ball is put back into play by a throw-in from the side line by a member of the opposing team. The ball is thrown in from the point where it crossed the side line as it went out of bounds. The ball must be thrown in with both hands using an over-the-head motion, keeping both feet on the ground.

(n) When the offensive team propels the ball across the defensive team's goal line but not in the goal, the defensive team is awarded a goal kick-a free kick taken from within its own goal area. The ball must come out of the penalty area to be in play.

(o) When the defensive team causes the ball to go behind its own goal line, excluding the portion between the goal posts, the offensive team is awarded a corner kick—a free kick taken by a member of the offensive team at the quarter circle, at the cornerflag post nearest to where the ball went behind the goal line. The flag post must not be removed.

(p) The game is started and, after a goal has been scored, is resumed by placing the ball in the center of the midfield line. Players must be on their side of the line until the ball is kicked. The ball must be kicked forward and must move at least 2 feet to be legal. The first kicker may not touch the ball twice in succession at the kick off. The opposing team must be 10 yards from the ball until it moves.

b.Speedball. Speedball offers vigorous and varied action with plenty of scoring opportunities. It is easy to learn and provides spontaneous fun. Little equipment is needed-a ball is all that is absolutely necessary. Speedball combines the kicking, trapping, and intercepting elements of soccer; the passing game of basketball; and the punting, drop kicking, and scoring pass of football.

(1) Place in the Program.Speedball, like soccer, should be introduced into the physical training program during the latter part of the toughening stage and used as a competitive activity in the sustaining stage. Speedball may be played any time weather permits but is primarily a spring or fall activity.

(2) Method of Play.The game is played as follows:

(a) Two teams of 11 each play the game under official rules, but any number of players may successfully constitute a team.An inflated leather ball, usually a soccer ball, is used. The playing field is a football field with a football goal post at each end. (See fig. 5-17.)

(b) The game starts with a soccer-type kickoff. The kicking team tries to retain possession of the ball and advance it toward the opposite goal by passing or kicking it. Running with the ball is not allowed, with the result that there is no tackling or interference. After the ball touches the ground, it cannot be picked up with the hands or caught on the bounce but must be played as in soccer until it is raised into the air directly from a kick. Then the hands may be used.

(c) When the ball goes out of bounds over the sidelines, it is given to a player of the opposing team and is put into play with a basketball throw- in. When it goes over the end line without a score, it is given to a player of the opposing team who may either pass or kick it onto the field.

(d) When two opposing players are contesting possession of a held ball, the official tosses the ball up between them as in basketball.

(e) Points are scored by kicking the ball under the crossbar of the goal post, drop kicking the ball over the crossbar, or completing a forward pass into the end zone for a touchdown.

Figure 5-17 is available in the illustration section at the end of the book.

(3) Basic Skills

(a) Shared Skills. Skills include kicking, passing, heading, and trapping (from soccer); punting, dropkicking and forward passing (from football); and passing, receiving, and pivoting (from basketball).

(b) Kickups. The kickup is a play in which a player lifts the ball into the air with the feet so as to legally play the ball with the hands. The kickup generally makes the transition from ground play to aerial play. The technique of making the play depends upon whether the ball is rolling or stationary. If the ball is rolling or bouncing toward the player, the foot is held on the ground with the toe down until the ball rolls onto the foot. Then the foot is raised, projecting the ball upward. If the ball is stationary, the player rolls it backward with one foot. Then the player places a foot where the ball will roll onto it. The player then raises the foot, projecting the ball as before. If a ball is rolling away, the player should stop it with a foot and play it as a stationary ball. The player can also raise the ball by standing over it with a foot on either side. The player presses the feet against the ball and jumps into the air, propelling the ball into the hands.

(4) Offensive and Defensive Play. The positions of players in speedball are much the same as in soccer. However, some of the positions have different names. Each team has 11 players. The forward line is composed of five players, the right end, right forward, center, left forward, and left end. The second line consists of right halfback, fullback, and left halfback. In the next line is the right guard and left guard. The player who defends the goal is the goal guard. The strategy employed in speedball during offensive play is very similar to that of soccer. There are two types of defensive formations in speedball: man-for-man and position defense. Man-for-man defense is recommended for beginning players.

(5) Abridged Rules

(a) Field. A speedball field is 100 yards long with additional 10-yard end zones on each end and 55 yards wide, the same as a regulation football field.

(b) Players. Eleven players are on a team. The goal guard has no special privileges.

(c) Time. Ten-minute quarters. Two minutes between quarters. Ten minutes between halves. Five minutes for extra overtime periods.(Begin the first overtime by a jump ball at center, same goals; change goals in the event of a second overtime period.)

(d) Winner of Toss. The winner of the toss has the choice of kicking, receiving, or defending a specific goal.

(e) Kickoff. A kick off starts the first and second half of a game and starts play after a score. The kick off is made from any place on the 50-yard line. The kicking team must be behind the ball when it is kicked. The receiving team must stay back of its restraining line (10 yards distance) until the ball is kicked. The ball must go toward the opposing team's goal before it can be played. Kick off out of bounds goes to the opponent at that spot. A kick off touched by the receiving team before going out of bounds, with no impetus added, still belongs to the receiving team. A kick off, in possession and control of the receiving team and then fumbled out of bounds, belongs to the kicking team at that spot.

(f) Starting Second and Fourth Quarters. The ball is given to the team that had possession at the end of the previous quarter, from out of bounds, as in basketball.

(g) Second Half. The first-half receiving team kicks off at the beginning of the second half.

(h) Fly and Ground Ball. The most characteristic playing rule of speedball is the difference between a fly ball (or aerial ball) and a ground ball. A player is not permitted to touch a ground ball with the hands and must play it as in soccer. A fly ball is one that has risen into the air directly from the foot of a player (for example, a punt, drop kick, place kick, or kickup).Such a ball may be caught with the hands provided the catch is made before the ball strikes the ground again.

(i) Kickup and Overhead Dribble. A kickup is a ball that is so kicked by a player that he can catch it himself. A bounce from the ground may not be touched with the hand because it has touched the ground since being kicked. This rule prohibits the ordinary basketball dribble, but one overhead dribble (throwing the ball into the air and advancing to catch it before it hits the ground) is permitted.

(j) Out of Bounds. If a team causes the ball to go out of bounds over the sidelines, a free throw-in (any style) is given to the opposing team. When the ball goes over the end line without scoring, it is given to the opponents who may pass or kick from out of bounds at that point.

(k) Tie Ball. In case two players are contesting the possession of a held ball, even in the end zone, a tie ball is declared. The ball is tossed up between them.

(6) Scoring

(a) Field Goal(3 points).A soccer-type kick, in which a ground ball is kicked under the crossbar and between the goal post from the field of play or end zone, is a field goal. (A punt going straight through is not a field goal for it is not a ground ball. The ball must hit the ground first.) A drop kick from the field of play that goes under a crossbar does not count as a field goal. A drop kick from the end zone that goes under the crossbar counts as a field goal;if it goes over the crossbar, it is ruled as a touch back, and the other team takes possession of the ball.

(b) Dropkick(2 points). A scoring drop kick must be made from the field of play and go over the crossbar and between the uprights. The ball must hit the ground before it is kicked (usually with the instep).

(c) End goal(1 point). This is a ground ball which receives its impetus (kicked or legally propelled by the body) from any player, offensive or defensive, in the end zone and which passes over the end line but not between the goal posts.

(d) Penalty Kick(1 point).A ball kicked from the penalty mark that goes between the goal posts and under the crossbar is a penalty kick. The penalty mark is placed directly in front of the goal at the center of the goal line.

(e) Touchdown(1 point). A touchdown is a forward pass from the field of play completed in the end zone. The player must be entirely in the end zone. If the player is on the goal line or has one foot in the field of play and the other in the end zone, the ball is declared out of bounds. If a forward pass is missed, the ball continues in play, but it must be returned to the field of play before another forward pass or drop kick may be made.

(7) Substitutions. Substitutions may be made any time when the ball is not in play. A player who is withdrawn may not return during that same period.

(8) Timeout. Three legal timeouts of 2 minutes each are permitted each team during the game.

(9) Fouls

(a) Personal Foul. Kicking, tripping, charging, pushing, holding, blocking, and unnecessary roughness of any kind, such as running into an opponent from behind, are personal fouls. Kicking at a fly ball and thereby kicking an opponent is a personal foul. Four personal fouls disqualify a player from the game.

(b) Technical Foul. Illegal substitution, more than three timeouts in a game, unsportsmanlike conduct, and unnecessarily delaying the game are technical fouls.

(c) **Violation.** Traveling with the ball, touching a ground ball with the hands or arms, double overhead dribble, violating tie ball, and kicking or kneeing a fly ball before catching it are violations.

(d) **Penalties.** The offended player shall attempt the kick.

(e) **Officiation of Fouls**

1 Fouls in the field of play allow no follow-up while fouls in the end zone always allow follow-up.

2 On penalty kicks, with no follow-up, only the kicker and goalie are involved.

3 On penalty kicks, with a follow-up, the kicking side is behind the ball and the defending side behind the end line or in the field of play. No one is allowed in the end zone or between the goal posts except the goal guard. The kicker must make an actual attempt at goal and cannot play the ball again until after another player plays it.

c. Volleyball. Volleyball is a popular sport. The game entails much physical activity, yet it is not strenuous. It is, therefore, a game for young and old alike, for beginners and for skilled players. It may be played indoors or outdoors on any type of terrain. As an informal activity, volleyball can be played by any number; as an organized activity, it provides, as few other sports do, a game for 12 to play in a limited area. While volleyball requires no great skill to play, it permits a high degree of proficiency. Players naturally get more enjoyment when they know the game and play it well. For this reason, instruction in the basic skills should be provided.

(1) **Place in the** Program Volleyball may be used occasionally as a competitive activity during the sustaining stage. Though it is a year- round sport, it should be included in the physical training program only when a more strenuous activity is impractical. Volleyball is an excellent off-duty activity.

(2) **Instruction.** Usually during the first three or four classes, a 10- to 15- minute period of instruction, followed by scrimmage, is enough to teach the basic skills, rules, and techniques of volleyball. More time, if available, can be given to teaching basic skills, but emphasize competitive play rather than formal instruction. Divide the unit into 6-person teams, organized on the basis of ability .Teams should be as nearly equal as possible.

(3) **Basic Skills**

(a) Passing the High Ball. The chest pass is the most effective method of playing the ball. To receive the ball, the knees are fixed, and the body is tilted forward. The elbows are raised sideward to a point in line with the shoulders. The wrists are extended in line with the forearm. The arms, wrists, and hands are rotated inward. To pass the ball, the hands are chest-high, thumbs pointing inward. The fingers are flexed, forming a cup, allowing them to contact the ball. On contact with the ball, the wrists are snapped while the fingers and elbows are pushed upward, sending the ball upward. A high ball is much easier to handle than a low one.

(b) Passing the Low Ball. A ball that is lower than the waist is one of the easiest to hit, but it is also a frequent cause of fouls (holding or carrying the ball). The best position for handling a low ball is to have the knees flexed and arms flexed at the elbows and rotated so the thumbs are pointing outward, the palms up. When the fingers contact the ball, the entire body reacts in a lifting motion. The arms and hands swing upward in a scooping action. It is important that the fingers, not the palms, contact the ball, and that the ball is batted, not thrown.

(c) Underhand Serve. Take a position behind the back line facing the net, left foot forward, holding the ball in the palm of the left hand. The left knee is flexed; the right knee is straight. Swing the right arm back and at the same time move the left hand (holding the ball) across the body in line with the right hip. Then swing the right arm forward hitting the ball off of the left hand with the palm of the right hand, raising the hips and arching the back in the same motion. Be certain to swing the right arm in a straight line, or the ball will be difficult to control. When the opposition is in formation, the server should try to place the ball in the right or left back area, and not near the net.

(d) Setup. A setup is a ball hit into the air near the net by one player, so that a teammate may hit or "spike" it sharply downward into the opponent's court. The chest pass is the best pass to use. The ball is sent approximately 10 feet into the air toward the spiker so it will descend from 4 to 20 inches from the net.

(e) Spiking. The spike is a leap into the air and a sharp downward hitting of the ball into the opponent's court. A spiker must be able to spring easily from the floor, judge the movement of the ball, and strike it with a downward movement of the arm. To jump from the floor, step off with one foot and jump with the other. Stand with the right or left side to the net, facing the setup player. Much depends upon the setup player to place the ball in the proper position. The spiker jumps into the air and strikes the ball above its center so as to drive it downward. A snapping movement of the arm and wrist will drive the ball forward and downward with power and control. Aim for a weak spot in the opponent's defense.

(f) Blocking. The block is a defense technique used to prevent a spiker from driving the ball across the net. It is an attempt by one or more defensive players at the net to block a hard-hit shot by using the force of the ball to send it immediately back into the opponent's court. In an effective block, forwards on the defensive team spring into the air at the time of the spike, placing both hands and arms in the expected path of the ball. An effective block tends to upset the offense and presents another element for the spiker to worry about. To be effective, the blocker must anticipate the path of the ball and time the block with the spike.

(4) Abridged Rules

(a) The volleyball court is 30 feet wide by 60 feet long. (See fig. 5-18.)

(b) The top of the net is 8 feet high.

(c) A volleyball team consists of 6 players.

(d) A match consists of the best two out of three games.

(e) The first team scoring 15 points wins the game, provided that they have 2 points more than their opponents.

(f) A deuce game is a game in which both teams score 14 points. The game is continued until one team obtains a 2-point advantage over the other.

(g) Only the serving team can score. If the serving team commits a fault, it loses the serve to the opposing team.

(h) The team members who receive the served ball rotate one position in a clockwise direction.

(i) The ball is put into play by serving from behind the back line.

(j) A served ball touching the net results in the loss of the serve. At any other time during play, a ball touching the net is still in play.

Figure 5-18 is available in the illustration section at the end of the book.

(k) The ball is out of play when it touches the ground or goes outside one of the boundary lines.

(l) All line balls are good.

(m) The players must hit or bat the ball; they may not throw, lift, or grasp it.

(n) A player may not touch the ball with any part of the body below the knees.

(o) A player may not play the ball twice in succession. In receiving a hard-driven spike, a defensive player may make several contacts with the ball even if they are not simultaneous. All such contacts, however, must constitute one continuous play, and all must be above the knees.

(p) The ball may be touched no more than three times on one side of the net before being returned across the net to the opposing team.

(q) A player must not touch or reach across the net.

(r) A player may touch the line under the net, but may not cross it.

Chapter 6
EVALUATION OF PERFORMANCE DURING TRAINING

6001.OBJECTIVES OF COMBAT READINESS

The objectives of the Marine Corps physical fitness program are-

a. To contribute to the health and well-being of every Marine through regular exercise and health education.

b. To develop Marines who are physically capable of performing their duties in garrison and in combat.

c. To develop Marines with a reserve level of physical fitness that will help them win in combat.

d. To provide a medium for developing the self-confidence of the individual Marine and thereby enhancing overall discipline, morale, esprit de corps, unit efficiency, and the desire to excel within the Marine Corps.

6002. THE PHYSICAL FITNESS TEST

The physical fitness test (PFT) has distracted attention from the proper goal of the physical fitness program. Simply stated, the goal of the physical fitness program is the success of Marines in combat. Unit commanders must not allow the PFT to become an end in itself. The PFT should be used within the commander's program of combat readiness training as a standard, an index of the physical fitness of individual Marines at a given time and place. The PFT is not difficult, and it must be kept in perspective. All athletes should be first class, and first class is attainable for all Marines. The conduct of the PFT is described in Marine Corps Order 6100.3_.

6003. OTHER METHODS OF EVALUATION

The commander should use other ways of evaluation to provide variety and imagination. Intangible objectives, such as confidence and aggressiveness, are hard to measure. The use of either inspection or observation is necessary to evaluate objectives of this nature. The commander has several other methods available when evaluating the physical condition of the command. The additional methods of inspection, observation, medical examination, and testing are summarized as follows:

a. Formal Inspections. Formal inspections, using inspection officers and standardized rating criteria, may help evaluate unit physical fitness.

b. Daily Observation. Routine observation of physical performance and appearance can serve as an indicator of an individual's or unit's physical readiness. However, mere observation is not a totally reliable or accurate means of evaluation.

c. **Physical Examination.** Medical examination may detect any individual disability or detrimental physical condition. It may guide in application of remedial, therapeutic, or limited exercise programs.

d. Other Physical Fitness Tests. Some other tests are described in detail in this manual for information purposes. Commanders are cautioned, however, not to place too great an emphasis on testing.

6004. RESPONSIBILITY

The commander is responsible for the physical fitness of the command and for the measurement and evaluation of its physical readiness. The goal of combat readiness training remains to prepare Marines and units for the physical demands of combat. The bulk of the commander's combat readiness training program should be such activities as obstacle and confidence courses, endurance courses, and progressive load-bearing marches which will prepare Marines for combat. Group activities which are competitive and combative in nature are advantageous and add imagination and enthusiasm to the program. Activities such as orienteering, which is not only physically demanding but also competitive and which teaches essential subject skills, are excellent and motivate all concerned. A commander's goal should be to provide an integrated training program that is so aggressive and demanding of Marines and leaders alike that a PFT administered at any time will reflect that each and every Marine is both physically and mentally prepared for the demands of combat.

6005. COMMANDER'S PHYSICAL FITNESS PROGRAM

The Commandant of the Marine Corps has directed that the physical fitness test shall be the universal measure of adequate individual physical fitness. The focus of a commander's physical fitness training program, however, should not be to prepare Marines to pass the physical fitness test, but to prepare them for the physical rigors of combat.

a. Administration. The physical fitness program requires each Marine to spend a minimum of 3 hours per week in physical fitness training, to be tested semiannually, and to obtain a minimum standard of third class. The attainment of a higher score is a laudable individual goal which should be encouraged but should not become a unit objective. Overemphasis could be detrimental to training required to develop the complete Marine. All commanders should exercise sound judgment as well as positive, aggressive leadership in striving toward the training objective.

b. Variety of Activities. The program must include the following elements:

(1) Physical Readiness Training. physical readiness training is a complete physical training program which develops and maintains the strength, endurance, and physical skills needed to sustain the individual during combat.

(2) Remedial Physical Conditioning. Remedial physical conditioning is a process by which physically substandard Marines are conditioned to meet prescribed standards.

(3) Competitive Conditioning Activities. Competitive conditioning activities consist of teams or individuals competing against an opponent to win. This involves a combination of sports and military skills designed to foster competitive attitudes and develop unit pride and esprit de corps.

(4) Occupational Conditioning. Marching long distances, dry net training, etc., should not be ignored as a means to fulfill the time requirement.

c.Time. In order to develop the desired level of physical fitness, unit commanders are advised that a minimum program of 3 hours per week is mandatory. The program must encompass at least three exercise periods weekly. Each period should include calisthenics, running, or other forms of vigorous activity .The minimum desirable length for each exercise period is 60 minutes. Shorter periods may be authorized by the unit commander if dictated by the local training situation. Unit commanders are encouraged to use the normal working day to satisfy this training requirement. However, commanders are authorized to conduct required physical fitness training during off-duty hours when the mission, workload, personnel status or other significant factors preclude fulfilling it during duty hours.

d.Action. Marine Corps Order 6100.3_ requires that commanders-

o Establish and maintain an effective physical fitness program which ensures that all Marines maintain an acceptable level of physical fitness.

o Ensure that all individuals are medically qualified prior to participating in the physical fitness program.

o Establish a minimum physical fitness program of at least 3 hours per week. Physical training may be authorized on an individual basis at the discretion of the unit commander.

o Ensure that all Marines participate in physical fitness conditioning activities commensurate with their medical qualifications and limitations.

o Conduct physical fitness testing for all Marines.

o Place all Marines who fail to pass the PFT on a daily command-supervised remedial physical conditioning program until they pass the test.

o Ensure that results of physical fitness testing are entered on section A of the fitness report. Comments are placed in section C in connection with attainment of superior physical performance or a medical excuse from the PFT.

6006. OTHER FORMAL TESTS

Commanders must realize that the best index of combat readiness is personal observation of performance on conditioning marches of 15 miles while under load. This manual presents three additional indicators of unit fitness. These indices are the physical readiness test, which was formerly the Marine Corps standard and is now routinely administered at Officer Candidate School; the airborne trainee physical fitness test which many Marines who attend U.S. Army formal schools are required to take; and the battle fitness test, the ultimate test of the combat readiness of Marines. Commanders may wish to include these tests or portions thereof within their combat readiness programs.

6007. PHYSICAL READINESS TEST

a. Purpose. The purpose of this test is to measure the Marine's ability to meet the minimum standard of physical readiness for duty in the field. This test is currently administered to all male officer candidates at Officer Candidate School, MCCDC, Quantico, VA. Female candidates do not take this test.

b. Test Area. Administration of this test requires an area suitable for crawling, jumping, and running 160 yards to include a trench that is 8 feet wide.

c. Equipment. The equipment required for this test includes—
o A platform, rail on a ledge, or a box 18 inches high,
o A 20-foot climbing rope,
o A stopwatch.

d. Conduct of Test. Adequate timers and other supervisory personnel must be provided by the company staff. All events should be conducted in a single session of one morning or afternoon. The events may be run in any sequence.

e. Uniform. The uniform for the physical readiness test is as follows: full utilities, cartridge belt, belt suspender straps, two canteens (full of water), rifle, and helmet. Depending on the heat condition, the PT shirt may be worn in place of the utility jacket. All equipment will remain secure throughout the test. Except for emergencies, canteens are not to be emptied.

f. Events
(1) Event 1: Climbing Uphill (Step-ups)This event simulates marching uphill at a rapid and steady pace. The Marine stands in front of the step. On the command GO, the Marine places either foot on the step and steps up. Hand pressure on the knee may be used if desired. The Marine then stands erect and steps backward and down, on one foot at a time. The same leg may be used for each repetition or the legs may be alternated. For 100 points, Marines will complete 60 repetitions in the maximum allotted time of 1 minute and 20 seconds (80 seconds).
*(2) Event 2: Rope Climb. T*his event resembles entering and leaving a hovering helicopter, using ropes in house-to-house fighting, and ascending and descending landing nets. The Marine assumes a sitting position on the ground at the bottom of the rope with hands at the highest point that can be reached. On the command GO, the Marine jumps up and climbs to the top. The Marine must touch the 20-foot mark with one hand, then descend without sliding or dropping. Marines must reach the 20-foot mark in the maximum allotted time of 30 seconds or less.
(3) Event 3: Evacuation. This event simulates reaching and evacuating a wounded Marine under fire. The Marine assumes the prone position at a distance of 50 yards from a casualty who is of approximately equal weight. On the command GO, the Marine springs up and covers the distance in a sprint. The Marine lifts the casualty to a fireman's carry and returns to the starting point. Assistance by the casualty is permitted. The tested individual must carry all equipment; the casualty has none. Marines must negotiate the course in a maximum time of 47 seconds or less.
(4) Event 4.'Advance by Fire and Maneuven. This event simulates advancing and assaulting during an attack. The Marine assumes the prone position at the starting point. On the command GO, the Marine then alternately crawls, runs in a zigzag manner, rolls, jumps, and assumes different firing positions for the next 130yards. Before reaching the objective, the Marine must "hit the deck," roll over, and assume the firing position three times. While assuming the firing position, the Marine must pull back the charging handle, take aim, and squeeze the trigger before rising from the ground. After rising from the ground for the third time,the Marine continues running and jumps to clear the 8-foot trench in one leap. The Marine then finishes the course in

the maximum allotted time of 2 minutes. The Marine is allowed multiple attempts only within the specified time limit.

(5) Event 5: Forced March. This event is a combat run of 3 miles without halts and against time. The Marine must double-time. Marines must complete the 3-mile course in a maximum allotted time of 30 minutes or less.

g. Event Standards

(1) To successfully pass the physical readiness test, the Marine must pass each event with a minimum of 80 points for each. Points are then assessed for each event in accordance with the scoring tables. (See fig. 6-1.)

(2) If a Marine fails one event, the highest possible score that the Marine may receive for the entire test is 79 percent. (See example 1 in fig. 6-2.) If two events are failed, the highest possible score for the entire test is 69 percent, regardless of the total points accumulated. (See example 2 in fig. 6-2.) If three events are failed, the highest possible score is 59 percent. Failure in four or more events will result in a score of zero for the entire test.

(3) To convert the total points for all five satisfactorily passed events to a final percentage score, total the accumulated points for all five events and divide by five. (See example 3 in fig. 6-2.)

Figure 6-1 is available in the illustration section at the end of the book.

Figure 6-2 is available in the illustration section at the end of the book.

6008. AIRBORNE TRAINEE PHYSICAL FITNESS TEST

a. Use and Composition of the Test.

(1) Use. The airborne trainee physical fitness test is a means of determining the physical ability of the applicant for acceptance to and retention in the airborne course of instruction.

(2) Test Events The test battery consists of five events as follows: chinups, knee bender, pushups, situps, and an endurance run.

b. Method of Scoring and Standards

(1) Scoring. The Marine will be scored by a trained scorer who is thoroughly familiar with the minimum standards for the test events. The Marine will be scored on a pass or fail basis. The performance on each event may be recorded on the reverse side of DA Form 705 (Physical Fitness Testing Record).The examinee records the personal information on the face of the card as specified on the first, second, and third lines. The face of the card is identified by writing diagonally across the lower half of the card "Airborne Trainee PPT—See reverse side."Using line 7 on the reverse side of the scorecard, the scorer enters the test title in the block entitled "(Other (Specify))" and completes the required information on the remainder of that line. The "Remarks" section can be used to record the test event titles and the applicant's performance on each test event.

(2) Standards. To successfully pass the test, the examinee must reach the standard in each test event. The standards follow:

(a) Chinups

(b) Knee bender 80(2-minute period)

(c) Pushups 45(2-minute period)

(d) Situps45(2-minute period)

(e) Endurance runs:2 miles in 15:54 minutes or less in athletic gear and 4 miles in 32 minutes in utilities and running shoes.

c. Uniform for Testing

(1) Examinees. The prescribed uniform for test participation is boots and the work uniform of the season. No headdress is worn. When climatic conditions permit, jackets or outer shirts may be removed.

(2) Officials. Scorers and other test officials should be uniformly and distinctively dressed for contrast with Marines being tested.

d. Test Administration Procedure

(1) Preparation for Test. The administration of the test to a large group makes it mandatory that the test effort be organized and efficiently operated. All testing is not completed with large test groups; on certain occasions, individuals and small groups are tested. Care must be exercised to administer the test uniformly and to standardize the conduct of all elements of the test. Regardless of the size of the test group, the following elements of sound test administration should be included:

(a) An orientation to include the purpose, method of administration, scoring of the test, preparation of the score card, and required standards.

(b) A correct demonstration of each event to ensure that there is no misunderstanding of the proper form and required standards.

(c) Completion of all five test events in one test period with all Marines taking the events in the same sequence.

(d) Adequate rest periods between the test events to allow for recovery before the next event.

(2) Method of Administration with a Large Group. With a minimum of 12 lanes per test event, 14 officials can administer the test battery to 150 or 200 Marines in 2 hours. The officials are designated as follows: one officer in charge, one demonstrator,and 12 scorers. If more or fewer Marines are to be tested, a greater or lesser number of officials will be required. The following procedure is recommended:

(a) Conduct an orientation and ensure the examinees have properly completed their score cards.

(b) Assign Marines to lanes and caution them to remain in the same lane order throughout the test.

(c) Explain and demonstrate the chinup event, administer it, and score it. Then proceed to the knee bender and pushup events, and administer them in the same manner.

(d) Grant a 5- to 10-minute rest period after the pushup event. Advise against excessive consumption of water during the break period.

(e) Explain and demonstrate the situp event, administer it, and score it as prescribed. Then move to the run area, explain the running event, and complete it.

(f) Retain the scorecards at the completion of the running event.

(3) Method of Administration with a Small Group. A similar procedure is followed for the testing of individuals and small groups. The informality usually associated with small groups must not conflict with sound test administration. With fewer examinees, fewer officials are required.

e. Description and Explanation of Test Events

(1) Test Event l:Chinups

(a) Purpose. This event is devised to test arm and shoulder flexor strength.

(b)Equipment There is one horizontal bar per lane, made of plumber's pipe or a gymnasium horizontal bar 1 1/2 inches in outside diameter. The bar must be rigidly supported at a height of 8 feet above the ground. The upright supports must be 5 feet apart. There must be a movable stand at each bar for short Marines to stand on to reach the bar.

(c) Officials There is one scorer per lane.

(d) Organization. The examinees, holding their scorecards, stand in order behind the restraining line in their respective lanes. The scorers take each scorecard when the Marine is called forward for the test.

(e) Starting Position The bar is grasped with the palms turned toward the face, the thumbs underneath the bar. The body is fully extended in a "dead" hanging position with the arms straight and the feet above the ground.

(f) Movement. Pull the body directly upward until the chin is placed over the bar. Lower the body until the elbows are completely straight and the body is again in the "dead" hanging position. Repeat as many times as required.

(g) Instructions. Explain and demonstrate the fully extended "dead" hanging position with the proper grasp. Show that the chin is placed over the bar at the top of the movement and that the arms are fully extended, the elbows completely straight, at the bottom of the movement (the hanging position). Explain that the body must be kept from swinging and that it is permissible to raise the legs and flex the hips when pulling up, but that any kicking, bicycling, or jerking motion with the trunk or legs is not acceptable. Inform the examinees that no penalty is exacted for hanging on the bar to rest in the bottom position but that this is not to their advantage. Tell them that half-completed chinups are not counted, and that the scorer will repeat the number of the last correct chinup when incorrect execution is detected.

(h) Administration and Scoring. Caution the examinees to assume the "dead" hanging position and wait for the scorer's command to begin. The scorer is at the examinee's left with a clear view of the bar. If the examinee begins to swing widely, the scorer should stop the swinging by extending the left arm across the front of the examinee's body, being sure not to hinder the execution of the chinup. The scorer counts aloud the number of chinups correctly executed. When a chinup is not correctly executed, the scorer repeats the number of the last correct one. The scorer records the number of correct chinups on the scorecard and returns the card to the examinee.

(2) Test Event 2:Knee Bender

(a) Purpose. This event measures the strength and endurance of the leg muscles.

(b) Equipment. None.

(c) Officials There is one scorer per lane.

(d) Organization. Marines stand in numerical order behind the restraining line in their respective lanes. The scorer takes each scorecard when the Marine is called forward for the test.

(e) Starting Position. The feet are spread less than shoulder width apart, hands on hips, thumbs in the small of the back, elbows back.

(f) Movement. Do a knee bend and at the same time bend slightly forward at the waist and thrust the arms between the legs until the extended fingers touch the ground. The hands are about 6 inches apart. The bend is approximately a three-quarters bend. From this knee bend position, recover to the starting position by moving the body upward, straightening the knees, and returning the hands to the hips. Repeat as many times as required.

(g) Instructions. Explain and demonstrate the correct starting position. Be certain examinees understand the correct knee bend and that only the tips of the fingers touch the ground. Tell them the scorer will repeat the number of the last correct knee bender when incorrect execution is detected. Some of the common errors are failure to correctly bend the knees, failure to touch the ground, and failure to assume the completely erect position after the bend has been executed.

(h) Administration and Scoring. The scorer stands to one side so as to see that the knees are properly bent and the fingers touch the ground as prescribed. From this position, the scorer can view the examinee to see that a properly erect position is assumed after each knee bend. The scorer counts aloud the number of correctly executed knee bends. When a knee bend is done incorrectly, the scorer repeats the number of the last correct one. The scorer records the number of correct knee bends on the scorecard and returns it to the examinee.

(3) Test Event 3:Pushups

(a) Purpose. Pushups measure arm and shoulder extensor strength.

(b) Equipment. None.

(c) Official. There is one scorer per lane.

(d) Organization. Marines stand behind the restraining line in their respective lanes until the scorer calls on them to perform. The scorer takes each scorecard when the Marine comes forward.

(e) Starting Position. The front-leaning rest position is the starting position. The body is straight from head to heels, palms are flat on the ground directly underneath the shoulders, and elbows are straight and locked. The body weight is supported on the hands and toes throughout the event.

(f) Movement. Bending only the elbows, lower the body in one straight plane until the chest touches the scorer's hand. Straightening and locking the elbows, raise the body in one straight plane, returning to the original front-leaning rest position. Repeat as many times as required, keeping the body in a straight line from head to heels.

(g) Instructions. Explain and demonstrate that the arms are straight at the beginning and completion of the movement and that the chest must touch the scorer's hand, but the stomach and thighs must not touch the ground. Also explain that the whole body must be maintained in a straight line as it is lowered and raised; that is, there is to be no breaking at the hips or shoulders so that any body part is lowered or raised in advance of the other or as a separate segment .Dipping or rolling through the shoulders is illegal, as is lowering or raising the body with one arm or shoulder at any time. Resting is not permitted during repetitions. Instruct

Marines that the scorer will repeat the number of the last correct pushup when incorrect execution is detected.

(h) Administration and Scoring. It is recommended that Marines assume a prone position while placing their feet and hands in the proper positions. This permits them to rest while the scorer gets into position and, at the same time, provides a feel of the body in a straight plane from head to heels. The scorer lies on the right hip and side to the right of the examinee. This gives the scorer a clear view of the examinee's body to see any errors. The palm of the scorer's right hand rests flat on the ground underneath the lowest part of the examinee's chest. By keeping the right forearm flat on the ground at an angle in front of the examinee's right arm, the scorer's position will not prevent the examinee from lowering the body completely. The scorer's left hand is free to test the straightening of the elbow at the completion of the movements and to point out body segments being lowered or raised separately. When in position and ready, the scorer has the examinee assume the starting position and begin doing pushups. The scorer counts aloud the repetitions done correctly and repeats the number of the last correct pushup if one is done incorrectly. There is no penalty if the contour of the examinee's body causes the hips to protrude slightly out of line, provided that the whole body is raised and lowered simultaneously. The scorer enters the number of repetitions on the scorecard and returns it to the examinee.

(4) Test Event 4: Situps

(a) Purpose. This event primarily measures abdominal strength.

(b) Equipment. None.

(c) Official. There is one scorer per lane.

(d) Organization Marines stand behind the restraining line in their respective lanes until the scorer calls on them to perform. The scorer takes each scorecard when the Marine comes forward.

(e) Starting Position. The examinee lies flat on the back with knees flexed, both feet flat on the ground. The correct angle of the thighs to the ground is 45 degrees. If the heels are too near the buttocks, the applicant will not be able to sit up. Legs are spread shoulder width apart. The examinee interlaces fingers and places them behind the head in contact with the ground. The feet are not held by another person.

(f) Movement. Bend forward at the waist and raise the upper body until the head is directly over the knees. Heels are not to leave the ground. Elbows remain in the same plane to the head and body throughout the event. The upper body is slowly lowered to the starting position until the head touches the ground. Repetitions are done at a slow cadence with no rest periods.

(g) Instructions. Explain and demonstrate the correct starting position and the proper execution of the situps to be sure that Marines understand the movement. Warn them that their knees must remain flexed during each situp, the heels cannot leave the ground at any time, and they may not roll up on one side and push up with one elbow. Tell them they must do the repetitions at a slow cadence with no rest periods. Instruct Marines that the scorer will repeat the number of the last correct situp when incorrect execution is detected.

(h) Administration and Scoring. When the performer is in position and ready, the scorer has the performer assume the starting position and begin doing situps. The scorer counts aloud the correct executions. When a situp is improperly done, the scorer repeats the number of the last correct one. No situp is credited if the hands are unclasped from behind the head, if the back is used to bounce up from the ground (which means the shoulders would not touch the ground), or if one shoulder or elbow is used to push up. The examinee is not penalized if the heels slide forward slightly so long as the knees remain flexed and the heels maintain contact with the ground. The scorer enters the number of repetitions on the scorecard and returns it to the examinee.

(5) Test Event 5:Endurance Run

(a) Purpose. This event measures cardiovascular endurance.

(b) Equipment One stopwatch or watch with a sweep second hand.

(c) Area. A large training field on which a quarter-mile track has been staked out or a level road over flat terrain may be used as a running surface. A 1-mile route is designated with wooden stakes marking the start point, finish point, and one-quarter mile intervals.

(d) Officials For large groups, a scorer times the event and controls the conduct of the run, and a guide runs with the group and sets the pace.

(e) Organization The run is conducted with groups of Marines in a column formation. Company-sized units may run at the same time with platoons serving as running groups. The scorer issues the command to assume double-time.

(f) Starting Position. Marines are assembled in proper column formation (column of twos, threes, or fours, as appropriate to the size of the group) with short men to the front. When all is ready, the column is moved forward a short distance before the running period is started.

(g) Movement. At the command DOUBLE-TIME, MARCH, the examinees retain their places in the column formation and execute the command. Length of steps is about 40 inches. The scorer has the group execute the run. The formation is maintained during therun.

(h) Instructions. Marines are instructed to maintain formation while running and are informed that the guide will set the proper pace. They will be instructed in the command used to control the column in the execution of the test. The scorer should announce the 4-minute, 2-minute, 1-minute, and 30-second remaining time intervals.

(i) Administration and Scoring. The event may be administered as previously prescribed to a large group, to several Marines, or to an individual. An individual examinee usually does not require a guide or pacer. If the event is administered on a training field, the scorer may stand in the center as the runner(s) circles about the field. This method of administration relieves the scorer of running with each group to be tested. Scoring is based on successful completion of the run as prescribed. The scorer should announce the remaining times as prescribed in (h).

6009. BATTLE FITNESS TEST

The battle fitness test is a physical fitness evaluation which is currently under development as part of the basic warrior training concept plan.

Chapter 7
THE HUMAN BODY

7001.GENERAL

To implement an effective physical conditioning program, a leader must understand how the human body functions and how exercise affects the body .This chapter provides only an introductory discussion of anatomy and functioning.

a. Body Functioning During the Stages of Conditioning. For more on the stages of conditioning, see chapter 1.

(1) During the toughening stage, the waste products of muscle activity (lactic acids) collect more rapidly than the blood can remove them. This acid waste builds up in the muscle tissue and irritates the nerves in the muscle fiber, causing pain and stiffness. As the exercise program continues, more blood is carried through the muscle, removing the waste materials more rapidly and eventually causing the soreness to disappear.

(2) During the slow improvement stage, the blood circulation in the muscles increases, and the body as a whole becomes a more efficient machine. The improvement is rapid in the first few weeks, but as a higher level of skill and conditioning is reached, the improvement becomes less noticeable. The body reaches its maximum level of performance after 6 to 10 weeks and should then be maintained at this peak.

(3) During the sustaining stage, it is possible to maintain this state of conditioning through 15 to 20 minutes of exercise a day, but the exercise must be quite strenuous.

b. Diet. Regular exercise increases the appetite. If the desire for greater amounts of food is satisfied by a balanced diet, the body benefits.

(1) There are two main types of foods: body-building and energy- producing.

(a) Body-building foods consist of proteins, which build up tissue and repair wear and tear.

(b) Energy-producing foods include carbohydrates and fats. Carbohydrates provide a quick source of energy, while fats act as a reserve store of energy.

(2) In addition, food contains vitamins, mineral salts, and water. During hot weather and strenuous training periods, the body requires greater fluid intake.

(3) Diet should be supplemented with proper rest to allow the digestive system to digest the lood.

(4) Occasionally, especially during early stages of conditioning, strenuous exercise may cause vomiting. Although vomiting is not a frequent occurrence, it is not usually a cause for concern.

7002. SYSTEMS OF THE BODY

The systems of the body include the skeletal, muscular, circulatory, respiratory, endocrine, digestive, genitourinary, and nervous systems. Each has a different function, but all must work in cooperation with one another to insure a sound body. Of these systems, the first four are the most affected by exercise and are discussed in succeeding paragraphs.

7003. THE SKELETON

a. Bones. The skeleton (see fig. 7-1) is composed of about 206 bones, which-
o Provide structure for the body and support for the attachment of muscles,
o Protect vital organs, such as the brain, lungs, and heart,
o Manufacture red blood cells, which carry oxygen through the body.

Figure 7-1 is available in the illustration section at the end of the book.

b. Joints. A joint is a place of union between two or more bones. Joints can be-
o Immovable, such as in the face and head.
o Slightly movable, such as in the vertebrae or sternum.
o Freely movable, such as in the shoulder, hip, knee, ankle, and elbow. The bones in a freely movable joint are held in place by ligaments. Freely movable joints are of the greatest importance in physical training because exercise improves their mobility and stability.

c. Cartilage and Ligaments. The joints in the body, where bones connect, are supported by cartilage and ligaments.
o Cartilage is a tough, elastic, translucent tissue that acts as a shock absorber or buffer between bones. Examples are the discs between the vertebrae, the tissue attaching the ribs to the sternum, and buffers in the knee joints. Damaged cartilage does not heal.
o Ligaments are connective tissue that bind bones together. They contract and extend but are not elastic. Consequently, ligaments can be torn or strained. Damaged ligaments will heal to some extent.

7004. THE MUSCLES

a. General. Muscles are tissue, or an organ composed of tissue, which contract or extend to produce movement. Muscles are of three classifications: involuntary, voluntary, and cardiac. The voluntary and cardiac muscles are the most affected by physical training; physical training has little effect on involuntary muscles.(Seefigs. 7-2, 7-3, 7-4.)
o People have no control over the involuntary muscles, an example of which is the diaphragm,
o Voluntary muscles are the larger skeletal muscles which cause movement and which the individual can control.
o Cardiac muscle is found only in the heart and for all practical purposes is an involuntary muscle. Physical training has a significant effect on the fitness of cardiac muscles.

b. Muscle Structure. Muscle is composed of cells which are grouped into bundles called fibers. Groups of fibers form muscles. These bundles of fibers are held in place by sarcolemma, a thin, sheath-like material that surrounds the muscle bundles and secretes a fluid that lubricates the muscle tissue. The fused ends of the sarcolemma form the tendons which attach the muscles to the bones.

c. Attachment of Muscles. The arrangement of muscles on the skeleton provides the proper angle of pull to make movement possible. Voluntary muscles are usually attached to the skeleton in two places.

d.Action of Muscles. To produce motion and to do work a muscle usually shortens its fibers. The movement may be flexion such as bending the arm at the elbow, or extension, such as the straightening the arm. Some muscles raise the arms or legs, others depress the raised limb. Some muscles have the primary function of rotating the trunk from side to side. In addition, muscles act as stabilizers as well as prime movers.

Figure 7-2 is available in the illustration section at the end of the book.

Figure 7-3 is available in the illustration section at the end of the book.

Figure 7-4 is available in the illustration section at the end of the book.

7005. FUNCTIONING OF THE SKELETON AND MUSCLES

a. General. The skeleton and muscles function in close coordination. Muscles move the body, and bones support the muscles.

b. Effect of Exercise on Bones. Continuous exercise, particularly among younger people, usually strengthens the bones, allowing them to withstand greater stress and strain. Bones which are not used lose much of their strength, a factor which should be considered when individuals return to physical training after a prolonged period of inactivity. People in this category should refrain from activities which might result in bone breakage before the bone is exercised back to normal condition. The condition known as "march fracture" is usually a result of this condition.

c. Muscular Strength. When exercised vigorously to improve strength, muscle grows in size. In general, the larger the muscle, the stronger the muscle. Furthermore, trained muscle functions more efficiently than untrained muscle. Trained muscles contract more vigorously and with less effort. Regular and strenuous exercise also toughens muscle, making it firmer and able to stand more strain.

d. **Muscular Endurance.** Muscles naturally become fatigued during continuous, repeated work. Through specific exercises, such as chinups or situps, local muscle groups become exhausted long before an individual fatigues. Training to develop muscular endurance enables people to continue a relatively heavy load of exercise over a long period of time. Lengthened exercise periods improve muscular endurance.

e. Muscular Coordination. Improved speed and strength result in part from improved muscular coordination. An unskilled individual may use irrelevant muscles to perform a particular activity, increasing the amount of work without increasing the mechanical output. This increase in skill is highly desirable, but it should be offset by greater effort in duration or intensity to compensate for the loss in overload due to increased skill.

f. Muscular Fatigue. When the rate of work is increased, the energy required is proportionately much greater than the increase in rate. For example, if an individual doubles running speed, the amount of power demanded to do this is increased eight times.

g. Circulation in Muscles Regular, strenuous exercise causes the creation of new capillaries and the opening of inactive, latent capillaries, increasing blood circulation within the muscle by as much as 400 percent. This increases the supply of food materials and oxygen to the muscle, improving its endurance. This process takes about 8 to 12 weeks of regular conditioning in young adults and longer as age increases.

7006. THE CIRCULATORY AND RESPIRATORY SYSTEMS

a. The Circulatory System. The circulatory system transports blood to all parts of the body, removes waste products for disposal, and delivers protecting and repairing substances where needed. The circulatory system includes the heart and blood vessels.

(1) The Heart. The heart is a pump which forces blood through the blood vessels to the parts of the body. The heart is a little larger than a fist and is located in the left center of the chest between the lungs.

(2) Blood Vessels. The vessels carrying blood away from the heart are the arteries. They eventually divide into capillaries, the very small vessels through which diffusion and osmosis take place. The capillaries gradually increase in size, forming the veins which carry blood back to the heart.

b. The Respiratory System. The respiratory system performs the function of breathing, whereby oxygen is inhaled and carbon dioxide is exhaled. The respiratory system consists of the mouth, nose, trachea, lungs, and diaphragm.

o Trachea. The trachea, or windpipe, is a hollow, tube-like structure that carries air from the mouth to the lungs.

o Lungs. The lungs are elastic bags in the chest where the exchange of oxygen and carbon dioxide takes place.

o Diaphragm. The diaphragm is a thin, sheet-like muscle just below the lungs. During inhalation, the diaphragm flattens out and lowers, allowing the lungs to expand and fill with air. During exhalation, the diaphragm raises into a dome shape, helping to expel air from the lungs.

7007. CARDIOVASCULAR FUNCTIONING

a. General. Cardiovascular functioning is the combined functioning of the circulatory and respiratory systems. The chief organs of these systems, the heart and lungs, function together to provide oxygen to the body.(See fig. 7-5.)

Figure 7-5 is available in the illustration section at the end of the book.

b. Heart Action. The heart, the organ which propels blood through the blood vessels, is the chief organ of cardiovascular endurance. A person tires quickly if the heart lacks the capacity to circulate the blood through the body. The heart is a muscular organ and is developed by exercise just like any other muscle. The heart cannot be exercised alone; any exercise that puts a sufficient load on the heart, such as running or cycling, will also exercise other body parts. A conditioned heart tends to beat more slowly but pump more blood when at rest. This is known as an increase in "stroke volume,"a desirable condition because it enables the heart to pump more blood with a slower contraction rate.

c. Functions of the Lungs. Strenuous and regular exercise can improve the efficiency with which the lungs transmit oxygen to the blood by as much as 25 percent In a poorly conditioned person, some of the alveoli (air sacs) within the lungs are closed or collapsed. The forced breathing created by exercise over a period of weeks causes the air sacs to slowly expand, increasing the ability to absorb oxygen.

d. Relationship of Heart and Lungs. The heart pumps carbon dioxide-laden blood through the pulmonary artery to the lungs. In the lungs the carbon dioxide is exchanged for oxygen and the purified blood is returned to the heart by way of the pulmonary vein. The heart then pumps the blood through the aorta for circulation throughout the body. Blood moving into the muscles exchanges oxygen for carbon dioxide, after which time it is ready for the return trip to the heart.

e. Cardiovascular Functioning in High Altitudes. Marines to be employed in high altitudes should be acclimated in a similar area for 10 to 14 days prior to employment. Persons not accustomed to the rarefied air of higher altitudes tire more quickly and may collapse after rapid physical exertion. Air is much less dense at high altitudes than at sea level. Persons accustomed to sea level inhale only about 80 percent of the oxygen they are used to no matter how hard they breathe. Further, people accustomed to sea level or moderate altitude simply do not have enough red corpuscles in the blood to fulfill their needs at high altitudes. The red blood cells will increase over time.

f. Symptoms of Cardiovascular Malfunctioning. The brain is the first organ to react to a lack of oxygen. Unconsciousness results when the brain is denied sufficient oxygen. "Blacking out" is actually a defense mechanism to keep the body alive.

APPENDIX A
INSTRUCTOR HINTS AND INSTRUCTOR TRAINING

1. General. This appendix informs instructors on administration of exercise periods, commands, the extended rectangular and circle formations, methods of instruction and conduct of exercise activities, explanation of basic positions associated with the various activities, and instructor training.

2. Aim of the Physical Fitness Program. The aim of the physical fitness program is to prepare Marines physically for war. This can be conveniently split into two phases.

a. Preparation Phase

(1) Development of all-around physical fitness.

(2) Development of purposeful physical skills.

(3) Development of mental alertness.

(4) Development of character and leadership.

b. Maintenance Phase. Once the preparation is completed, the program must progress to a maintenance program. This phase revolves around the need to maintain the trained Marine's physical fitness for a particular role. This will include—

(1) Maintenance of a high standard of all-around physical fitness.

(2) Application of purposeful physical skills to combat skills.

(3) Development of leadership and self-confidence.

(4) Development of mental toughness and endurance.

3. Physical and Mental Development

a. Marines need to be developed physically and mentally. The physical training (PT) instructor must ensure that the program accomplishes this.

b. The system of the body functions best in response to progressive exercise. This means that Marines should master the easier exercises before they attempt the more difficult ones.

c. The body and mind are linked inseparably and must be considered as one in physical fitness training. Physical exercises assist in developing the mind.

d. The confidence in oneself and in one's comrades attained by physical achievement helps to promote the indomitable will to win. The Marine must be taught to realize this. It will help during arduous training, which at all times must be imaginative, realistic, and filled with enthusiasm. All physical fitness training must be associated with developing the qualities required of a Marine.

4. Planning and Preparation of ProgramsThe following factors must be considered when compiling a program (see also ch. 2)~

a. Aim or desired skill level.

b. Duration of the complete training program and number of physical training lessons.

c. Purpose of included exercises,drills, or practical applications.

d. Beginning/inventory fitness levels.

e. Medical facilities, medical personnel, and Marine Corps or local command regulations governing their proximity to the training.
f. Number of Marines to be trained.
g. Availability of instructors, assistants, and facilities.
h. Apparatus and equipment available.
i. Alternative training areas for varying weather conditions,
j. Stages of progression and tests of achievement.

5. Five Basic Principles of Program Design
a. Regularity. Regularity of exercise is far more important than the amount of exercise performed on a given day. A Marine should exercise a minimum of three times per week. Human muscle begins to atrophy after 3 days without exercise.
b. Balance. The program must be balanced in terms of total physical fitness. Imbalance is created when, for example, an individual merely runs. While the cardiovascular system is improved, general upper body strength may be lacking. Conversely, a person who just does weight training may develop strength, but not cardiovascular endurance.
c. Overload. If total body fitness is to be developed, then the muscular and cardiovascular systems must be made to perform more work than previously. This is done by increasing the frequency, intensity, and duration of the exercise. The body adapts to the amount of stress placed upon it and becomes stronger and more efficient.
d. Progression. To be successful, any type of training program must be progressive. This is an area where untrained persons conducting PT can cause injury to Marines resulting in failure of the PT program. Close supervision must be made of untrained instructors. The result of progressing too fast is that the class begins to reject PT because it is no longer enjoyable or motivating. The program should start at a slow pace, and the pace should increase as proficiency improves. Average persons will encounter three stages of progressiom the toughening stage, the slow improvement stage, and the sustaining stage. (See ch. 1.)
e. Variety. Some programs fail because the training becomes boring. Variety is essential and perhaps presents the greatest challenge to the commander and staff.

6. Grouping. Providing for different levels of physical fitness is particularly recommended in the early stages of conditioning. Older Marines and those in poorer physical condition should be expected to attain a group level of fitness, but they should be given more time to do it.
a. Homogeneous Grouping One simple method of providing for the difference in levels is to group them according to their condition. A two-group classification would divide Marines into highly conditioned and average groups. A finer classification could be obtained by dividing them into three groups—a highly conditioned, a moderately conditioned, and an unconditioned group.
b. Performance or Age Grouping. The segregation of Marines into different exercise groups should be based on physical fitness test scores or on the level of fitness they demonstrate. They may also be grouped at first according to age. A common classification by ages is under 30, 30 to 34, and 35 and above.

7. Preparatory Commands and Commands of Execution. The prepreparatory command describes and specifies what is required. The command of execution calls into action what has been prescribed. All preparatory commands are given with a rising inflection. The interval between commands is long enough to permit the average Marine to understand the first one before the second one is given.

8. Extended Rectangular Formation. The traditional formation for carrying on many physical training activities is the extended rectangular formation (fig. A-1). This formation is more compact than any other. It is the best type to use for large numbers of Marines because it is simple and easy to assume.

a. For the formation of oneplatoon, the base man paces off five paces from the stand, faces left and moves five paces, halts, and again faces left. With the base man positioned facing the stand, the platoon leader then commands. **FALL OUT AND FALL IN ON THE BASE MAN**. At this command, all Marines run to the designated area and reform. This procedure is preferred to marching the unit into position. If more control is desired, the unit may march at double time to the vicinity of the base man and then be directed to fall out and fall in on him. Much valuable time is wasted in the field by needless maneuvering of Marines at quick time in an effort to position the platoon or unit on the exact spot for the exercises.

b. A company-size unit assumes the extended rectangular formation from a column of threes or fours at normal intervals between squads. This extension can also be executed from a company mass without interval between platoons. In extending either a platoon or company-size unit, take your place at the head of the column and command:

(1) TAKE INTERVAL TO THE LEFT, MARCH. At this command, the Marines in the right flank file stand fast with arms extended sideward. All others turn to the left and run forward at double time. After taking a sufficient number of steps, all Marines face the front with both arms extended sideward. The distance between finger-tips is about 12 inches, and dress is right.

Figure A-1 is available in the illustration section at the end of the book.

(2) ARMS DOWNWARD, MOVE. At this command, the arms are lowered smartly to the sides.

(3) LEFT, FACE

(4) EXTEND TO THE LEFT, MARCH. At this command, those in the right flank file stand fast with arms extended sideward. All others turn to the left and run forward at double time. Spacing is the same as in (1) above, and dress is right.

(5) ARMS DOWNWARD, MOVE. Same as in (2) above.

(6) RIGHT, FACE.

(7) FROM FRONT TO REAR, COUNT OFF. At this command, the leading Marine in each column turns the head to the right rear, calls of **ONE** and faces the front. Successive Marines in each column call off in turn **TWO, THREE, FOUR, FIVE** in the same manner.

(8) EVEN NUMBERS TO THE LEFT, UNCOVER. At this command, each even-numbered Marine stride jumps sideward to the left, squarely in the center of the interval. In doing this, each one swings the left leg sideward, jumps from the right foot to the left foot, and smartly brings the right into position against the left.

c. To assemble the unit, you commandASSEMBLE TO THE RIGHT, MARCHAt this command, all return to their original position in the column at double time and reform on the base man.

d.It is recommended that the area for grounding equipment and arms be at the edge of, or well away from, the area to be used for exercising. To conserve time and ensure proper position of the unit, the base man or, if the unit is composed of several platoon-sizegroups, the various base men may precede the unit and establish their positions in relation to the instructor's stand.

9. Circle Formation.The circle formation is effective for conducting various exercise activities (fig. A-2). This formation has an advantage in that supervision of all Marines is facilitated and a moving formation is available which provides control. Guerrilla exercises, grass drills, and some forms of running are examples of activities which are more easily conducted in the circle formation than in the extended rectangular formation.

a. It is not advisable to have more than 60 Marines in a circle. When more must be accommodated, separate circles should be used.Concentric circles may be employed to reduce the size of the circle or to accommodate more persons. If concentric circles are formed, the different circles are made by designating squads for each circle.Each additional circle requires more Marines than the one inside it.For example, one squad of a platoon may form the inner circle and theremaimng three squads the outer circle.

b. When a platoon is to form a circle, the commands ai€IRCLEFORMATION, MARCH, FOLLOW MEUpon this command, the right flank squad of the column moves forward at double time with the leader of the platoon group gradually forming a circle in a counterclockwise direction. Each succeeding file falls in behind that on the right. After the rough outline of the circle is formed, the leader commandsPICK UP A **5**-YARD INTERVAL.

Figure A-2 is available in the illustration section at the end of the book.

c.The group may be halted and faced toward the center, or, if instruction is not necessary, the exercise activity may be executed without stopping the platoon.

lO.Leadership Techniques
a. Unless you experience all the exercises, you cannot appreciate how arduous they are, what movements are most strenuous and difficult, where the errors in performance are likely to occur, and what the proper cadence should be.

b. You must give everyone careful supervision and participate in the exercises to show that you can do them. When you participate, your assistant instructors should supervise because it is difficult for you to supervise and exercise at the same time.

c. Marines should never be kept too long in one position, especially a constrained one. They should never have to perform an exercise more times than they can do it without losing the proper form. You should insist upon proper form in the execution of all exercise activities.Even slight deviations from the proper form reduce the value of the exercise.

d. Avoid long explanations.As a rule, it should be necessary to give a full explanation of new exercises only. Explain the most essential features of an exercise first; add details later. Too many details at one time are more likely to confuse than to assist them.Minor corrections should be made to the entire class while the exercise is in progress (for exampHEADS **UP,** KNEES STRAIGHT). If necessary, follow this direction by the name of the person who is particularly at fault. If a Marine requires special attention, givethat person separate instruction by an assistant instructor to avoidwasting the time of the group.

e. The heavy demand on your voice can be lightened by training assistant instructors to assume some of the instruction. Using mass cadence is also an effective method of lessening the demand on your voice.

“Illustrations”

DAY	TIME	ACTIVITY
MONDAY	60 MIN	RUNNING ACTIVITY
TUESDAY	60 MIN	COMBATIVES
WEDNESDAY	60 MIN	CONDITIONING & LOG DRILLS
THURSDAY	60 MIN	CONDITIONING & RELAYS
FRIDAY	60 MIN	RUNNING & CONDITIONING
SATURDAY	60 MIN	RUNNING ACTIVITY

Figure 2-1. Sample A- -Single Activity.

DAY	TIME	ACTIVITY	REMARKS
MONDAY	20 MIN 20 MIN 20 MIN	RIFLE DRILL COMBATIVES RUNNING	PROGRESSIVE ACTIVITY SYSTEM
TUESDAY	30 MIN	GUERRILLA EXERCISES TABLE I	SINGLE ACTIVITY
WEDNESDAY	30 MIN	RIFLE DRILL	SINGLE ACTIVITY
THURSDAY	30 MIN	CIRCUIT INTERVAL TRAINING	SINGLE ACTIVITY
FRIDAY	20 MIN 20 MIN 20 MIN	STRENGTH CIRCUIT CONDITIONING DRILL CROSS-COUNTRY RUN	ROTATION ACTIVITY
SATURDAY	60 MIN	WARM UP RUN AND GRASS DRILL	SINGLE ACTIVITY

Figure 2-2. Sample B- -Various Time Periods, Activities, and Systems.

DAY	TIME	ACTIVITY
MONDAY	45 MIN	RUNNING – CROSS-COUNTRY
TUESDAY	75 MIN	TEAM ATHLETICS – TOUCH FOOTBALL
WEDNESDAY	45 MIN	RUNNING – CROSS-COUNTRY
THURSDAY	20 MIN 40 MIN	LOG DRILL DUAL COMBATIVES
FRIDAY	90 MIN	SPEED MARCH
SATURDAY	90 MIN	TEAM CONTESTS

Figure 2-3. Sample C--Sustaining Activities.

Route	Kilometers from SP	Recommended rate of march (kmph)	Remarks
SP: RJ 8th Div Rd—Superhighway	......	4	Bridge; hard surface; two guides.
RJ Jamestown—Hersey Rd	2,25	4	Traffic heavy; two guides; Jamestown Rd bears to right.
RJ Jamestown—Yankee Rd	7.10	4	Two guides
RJ Jamestown—Lightning Rd	10.50	4	Two guides
RJ Jamestown—Sunshine Rd	12.40	4	Light traffic; one guide Sunshine Rd; gravel, poor traction when wet.
RP; RJ Sunshine—Sedan Rd	14.80	4	Two guides

Figure 3-1.Example of Route Reconnaissance Report.

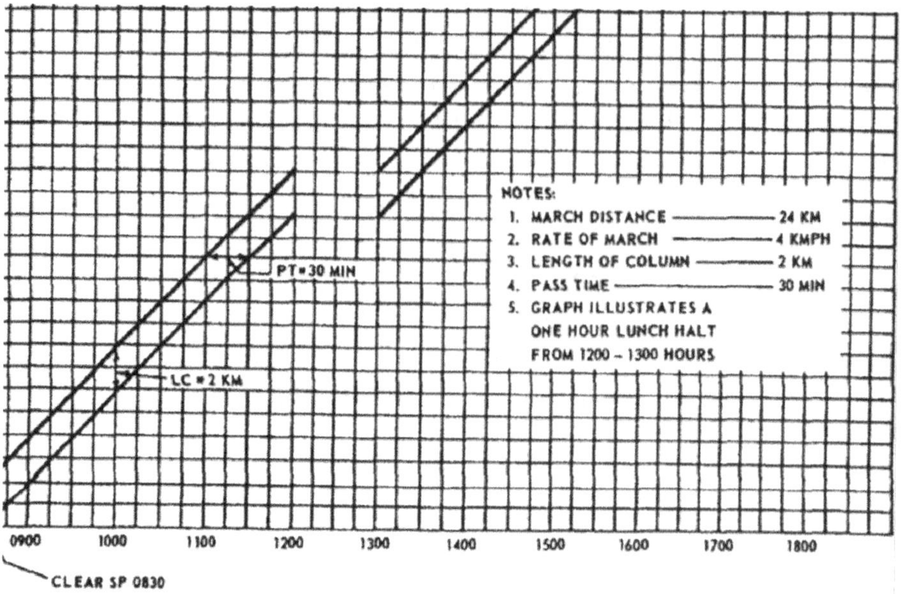

Figure 3-2.Road Movement Graph.

Figure 3-3.The Pace-Setter.

	SINGLE FILE	COLUMN OF TWO
2 m/MAN APART	24	12
5 m/MAN APART	54	27

GAP DISTANCES			
SERIAL:	(3-1) × 100 m	=	200 m
MARCH UNIT:	(12-1-2-) × 50 m	=	450 m
	TOTAL COLUMN GAP		650 m

Figure 3-4.Conditioning Drill One.

Figure 3-5.Conditioning Drill Two.

Figure 3-6.Conditioning Drill Three.

Figure 3-7. Rifle Drill.

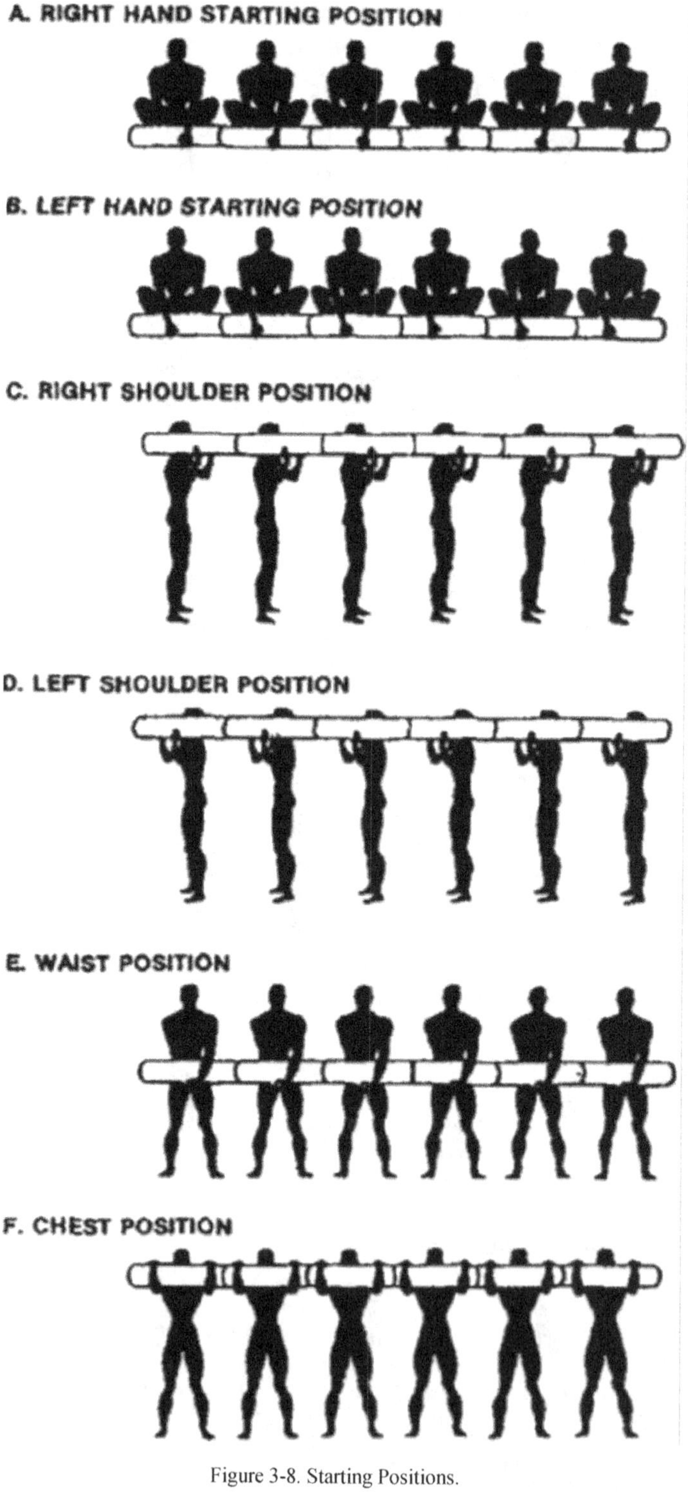

Figure 3-8. Starting Positions.

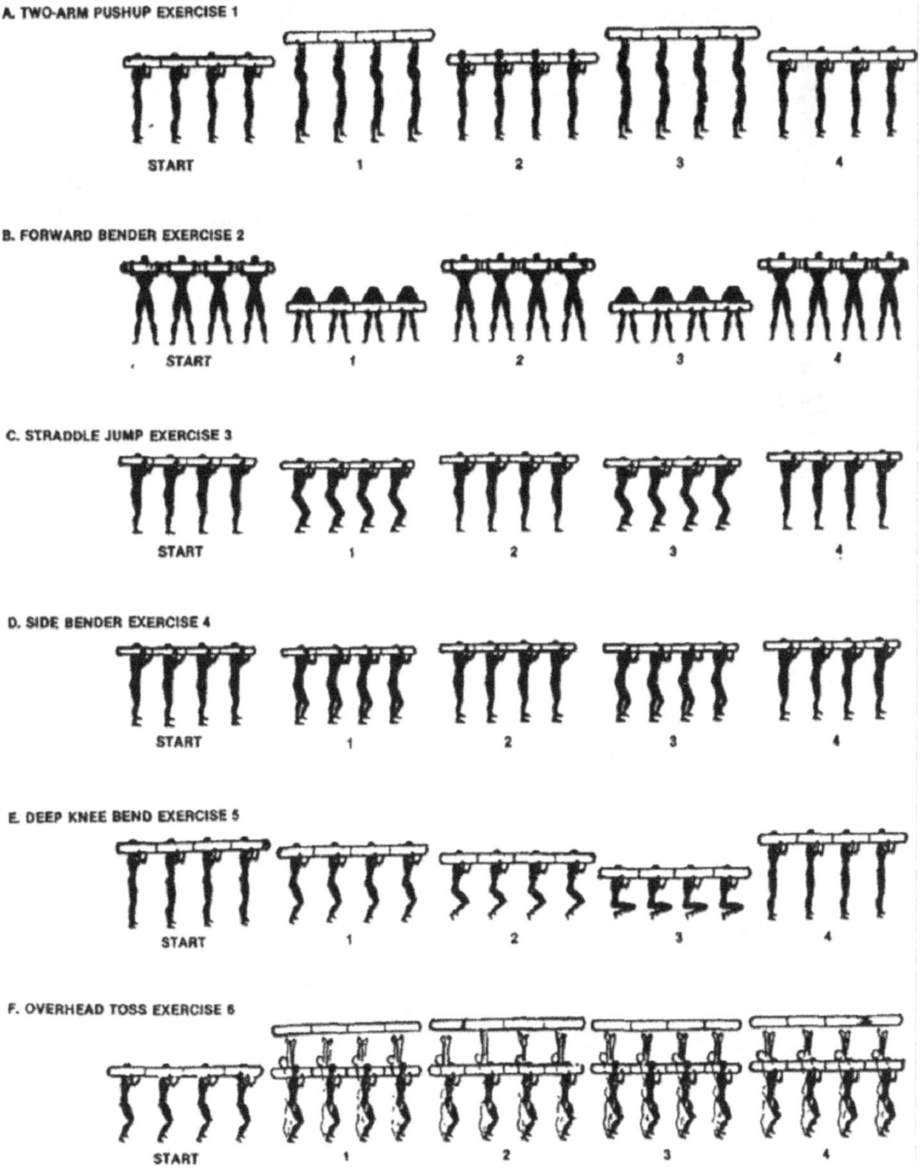

Figure 3-9. Log Drill

Figure 3-10.Basic Positions for Grass Drill.

Figure 3-11.Grass Drill One.

Figure 3-12.Grass Drill Two.

Figure 3-13.Guerrilla Table I.

Figure 3-14.Guerrilla Table II.

Figure 3-15.Proper Running Form.

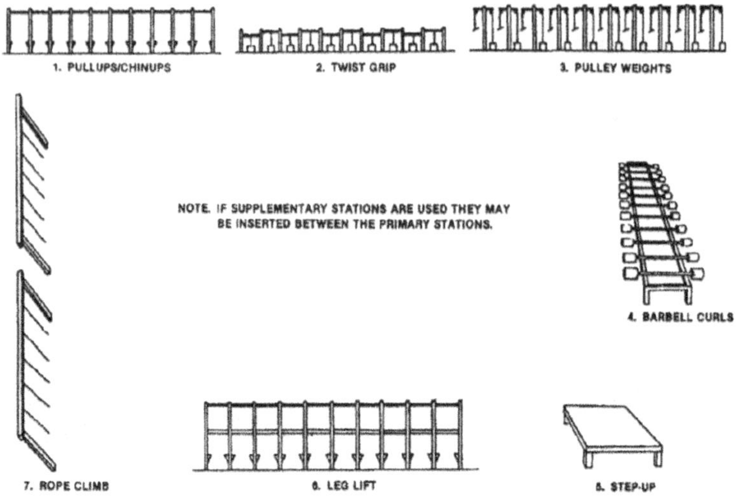

Figure 3-16. Fixed Strength Circuit.

STATION	ITEM	NO.	SPECIFICATION
1	BARBELL		1¼-INCH PIPE 5 FEET LONG WITH CONCRETE FILLED NO. 10 CANS
2	JUMP ROPE		¼- OR ⅜-INCH ROPE, 10 FEET LONG
3	TWIST GRIP		HANDLE 12 INCHES LONG, ROPE 4 FEET LONG, NO. 10 CAN CONCRETE FILLED.
4	INCLINE PLANE		¾-INCH PLYWOOD PLATFORM 2 FEET WIDE AND 6 FEET, 6 INCHES LONG, ELEVATED 10 INCHES AT ONE END. STRAP TO HOLD FEET DOWN.
5	WAR CLUB		HEAD IS 6 BY 12 INCHES, HANDLE IS 14 INCHES LONG BY 1¼ INCHES IN DIAMETER, ABOUT 20 POUNDS.
6	BICYCLE RIDE		PLYWOOD BOARD OR PLATFORM 2 BY 3 FEET WITH 2 BY 2 RUNNERS.
7	STEP – UP		A BOX OR STURDY PLATFORM 18 INCHES HIGH, 18 INCHES WIDE, 24 INCHES LONG.
8	ISOMETRIC PULL		TWO HANDLES 12 INCHES LONG WITH 4 FEET (BETWEEN HANDLES) OR LIGHT WIRE CABLE OR ¼-INCH ROPE

Figure 3-17.Movable Strength Circuit.

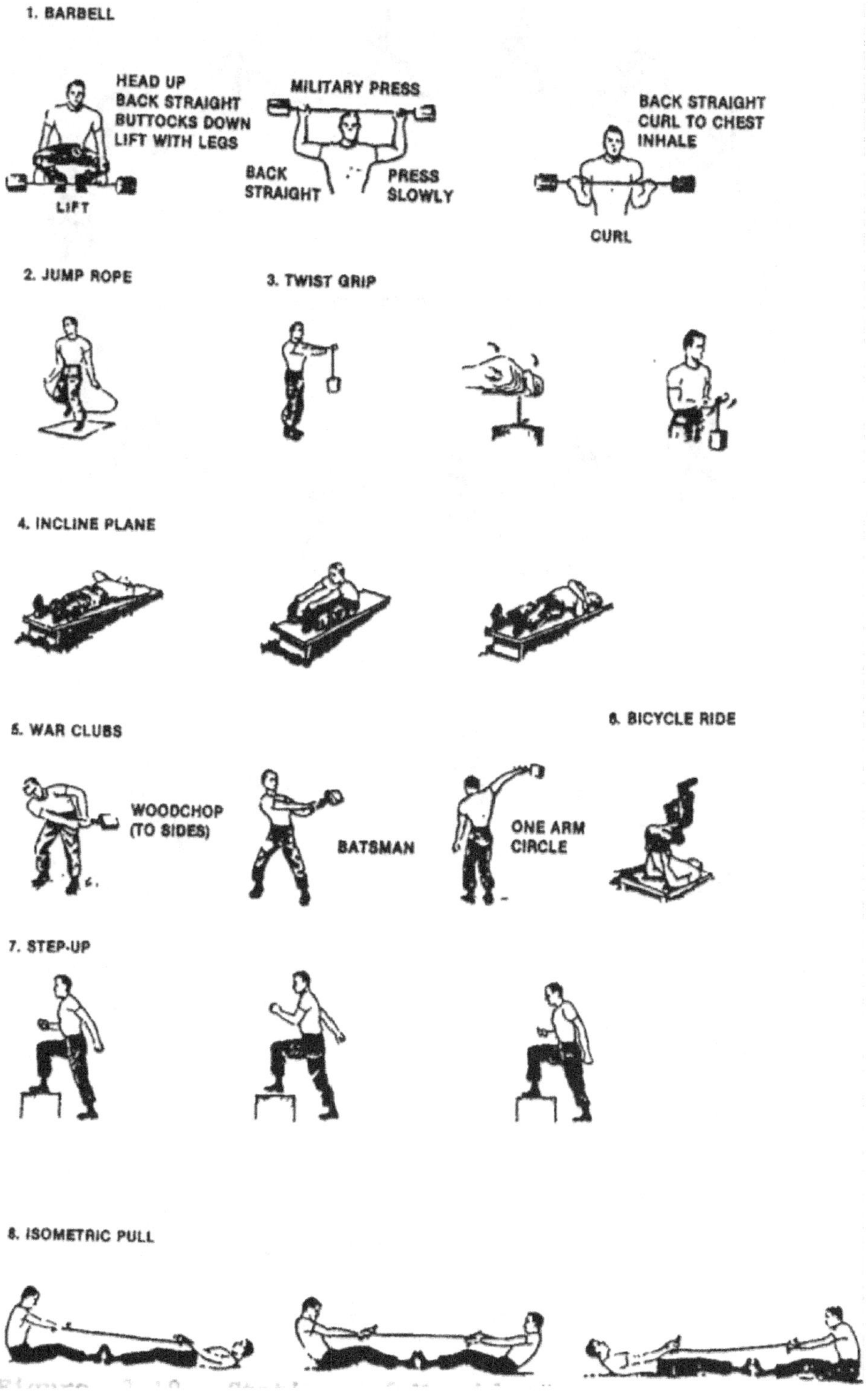

Figure. 3-18.Stations of Movable Strength Circuit.

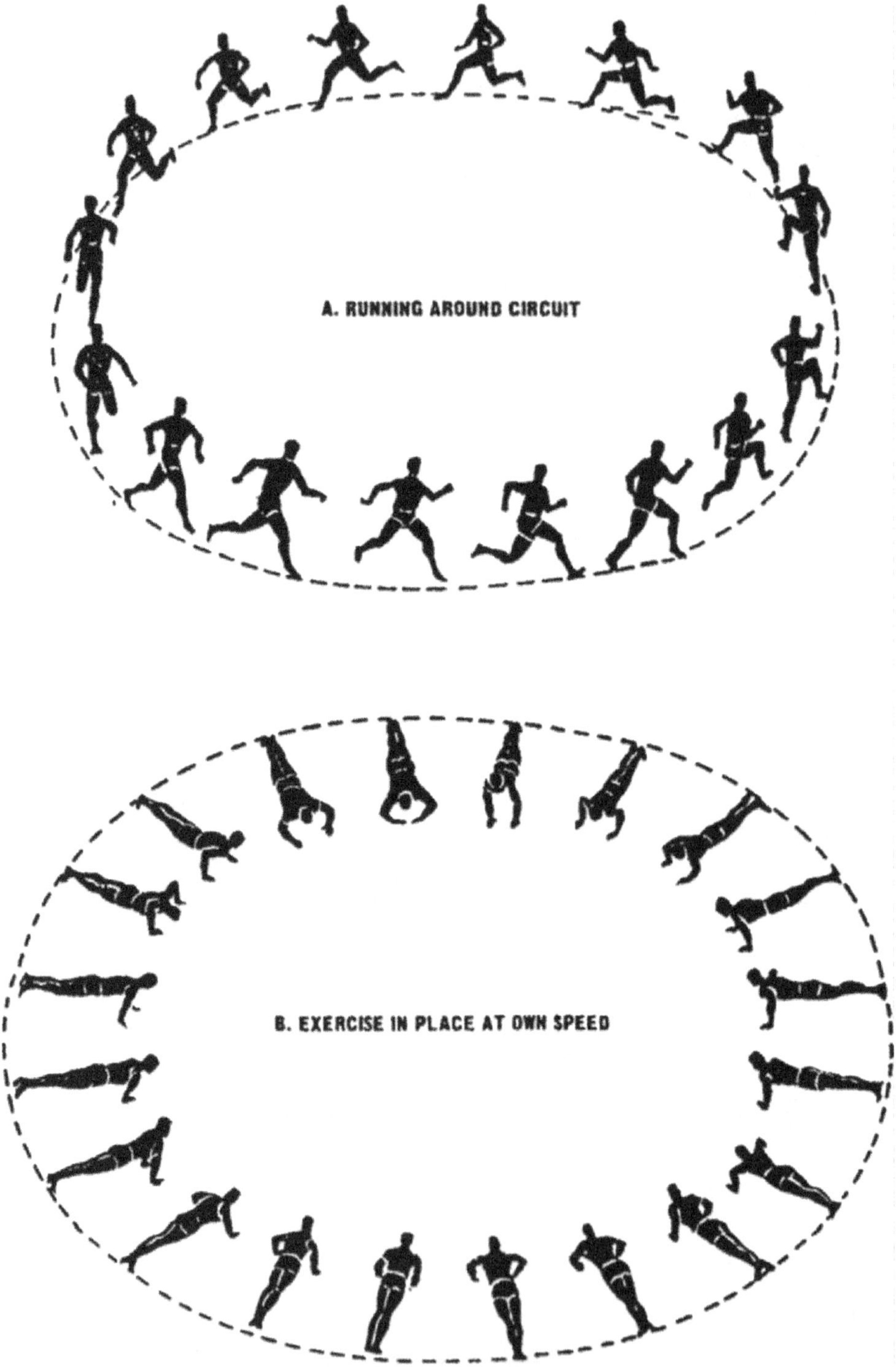

Figure 3-19. Circuit-IntervalTable.

Figure 3-20. Jumping.

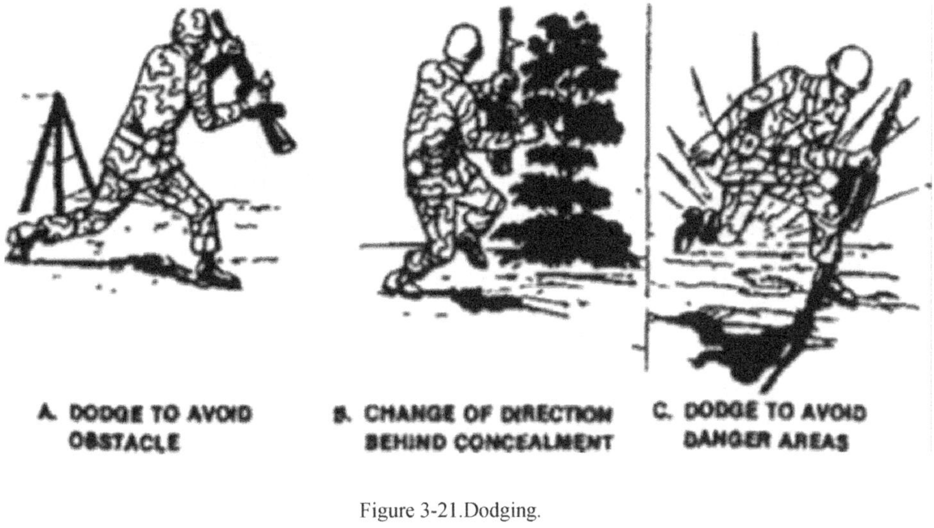

Figure 3-21.Dodging.

Figure 3-22.Climbing and Surmounting Vertical Objects.

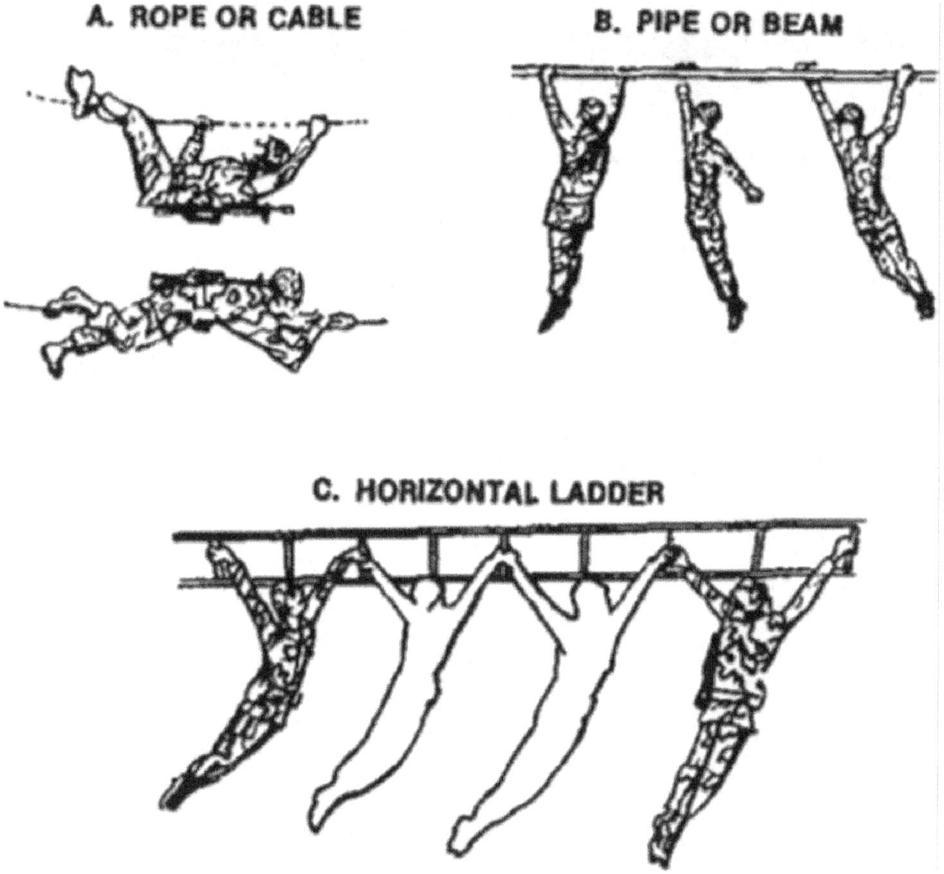

Figure 3-23. Traversing Horizontal Objects.

Figure 3-24.Crawling.

Figure 3-25.Throwing.

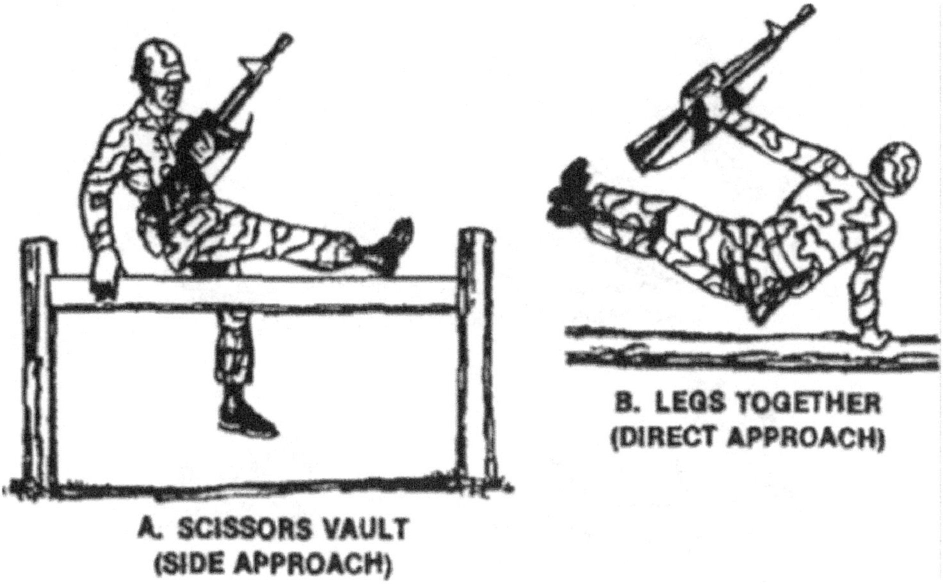

Figure 3-26. Vaulting

Figure 3-27.Balancing.

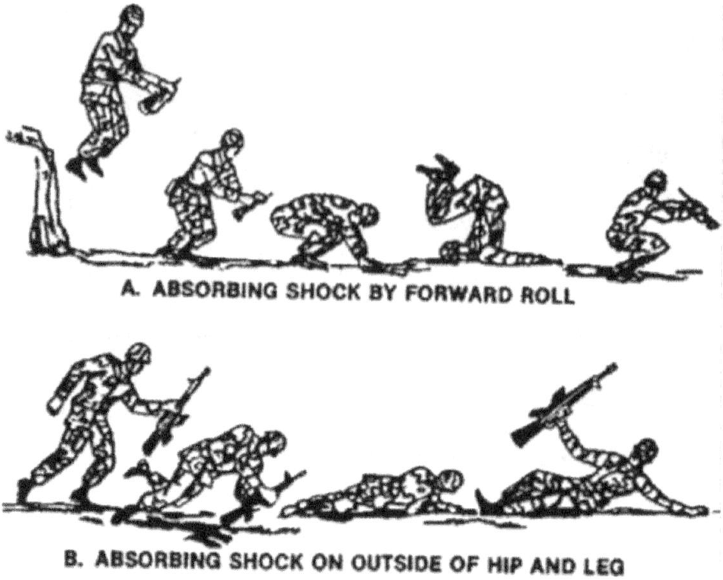

Figure 3-28.Falling.

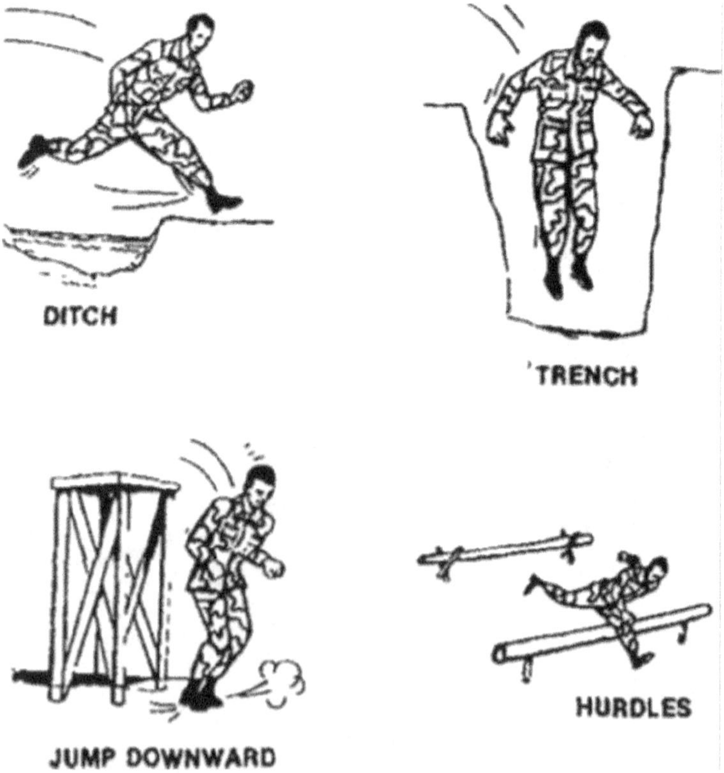

Figure 3-29.Jumping-Type Obstacles.

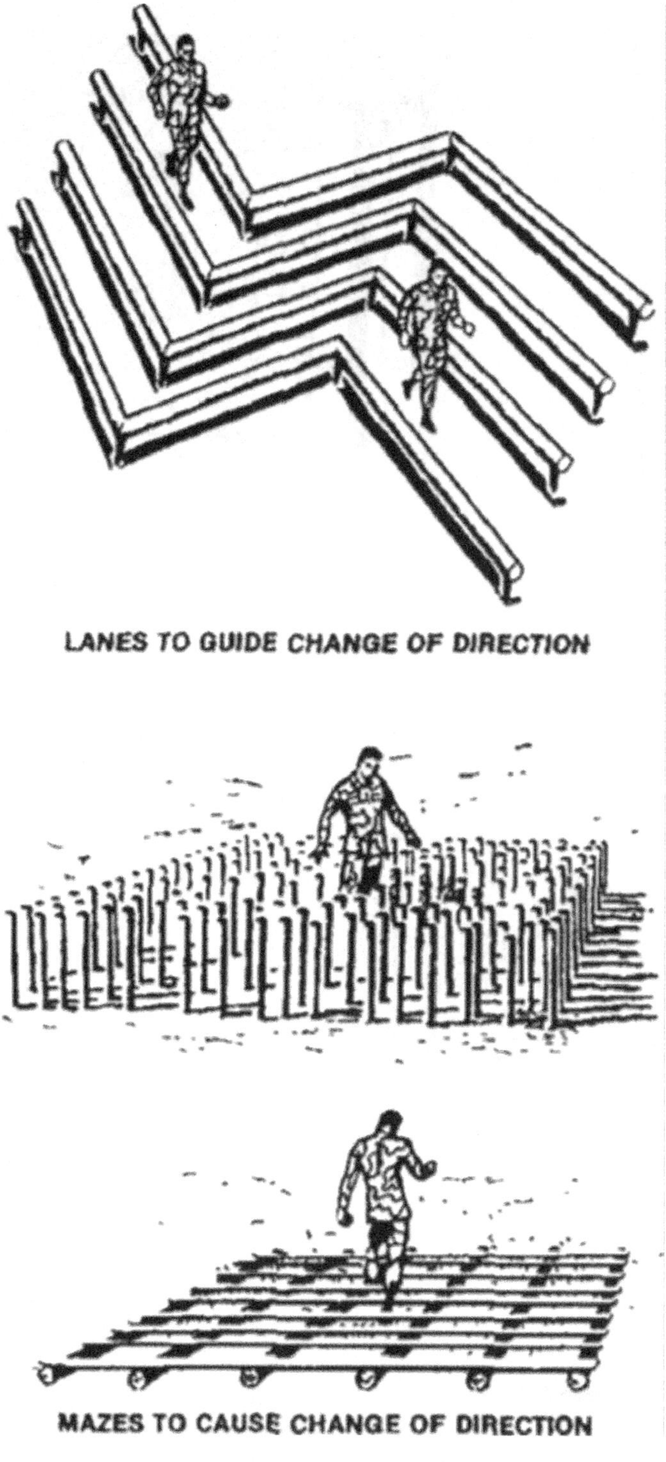

Figure 3-30.Dodging-Type Obstacles.

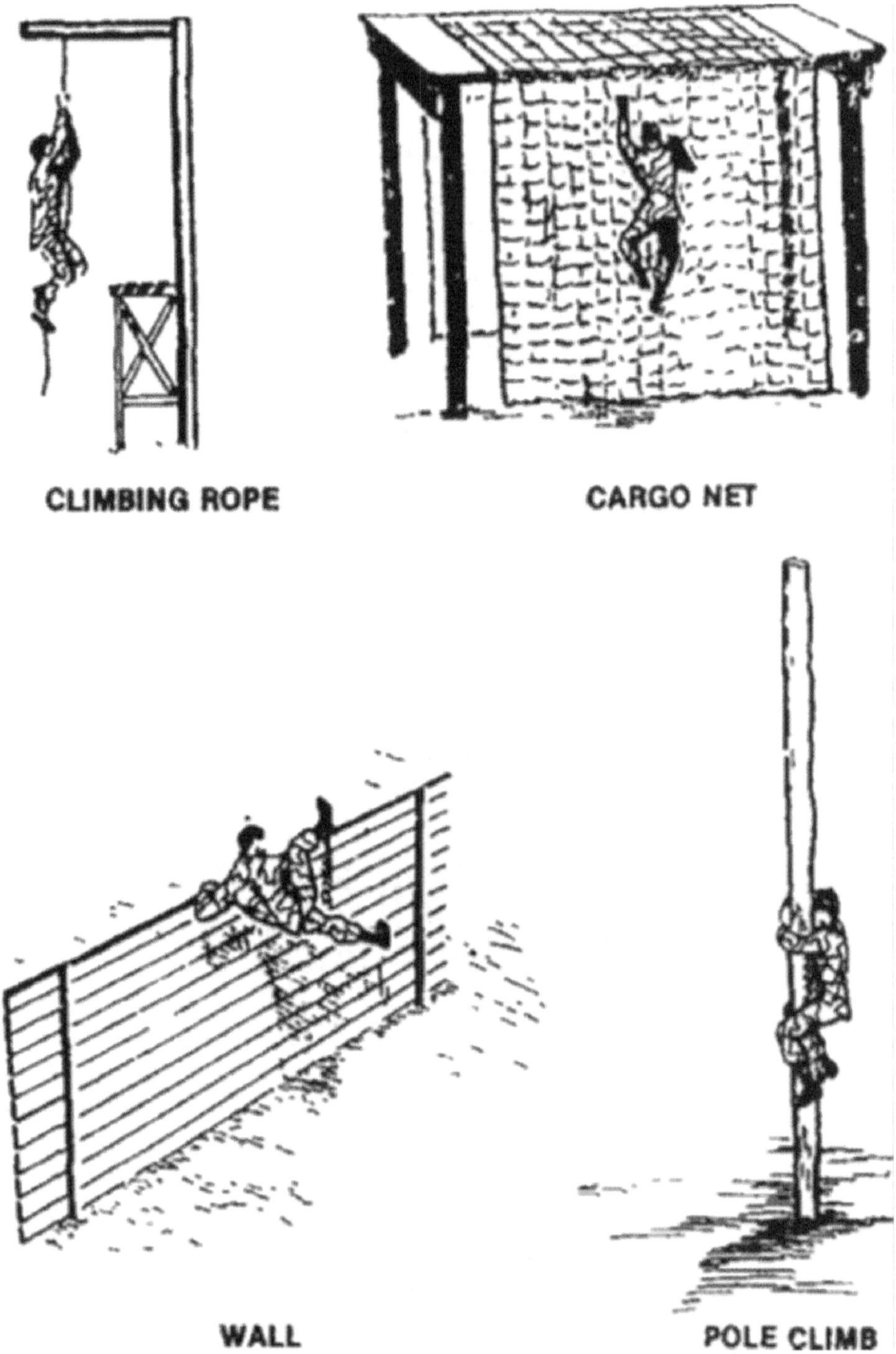

Figure 3-31. Vertical Climbing Obstacles.

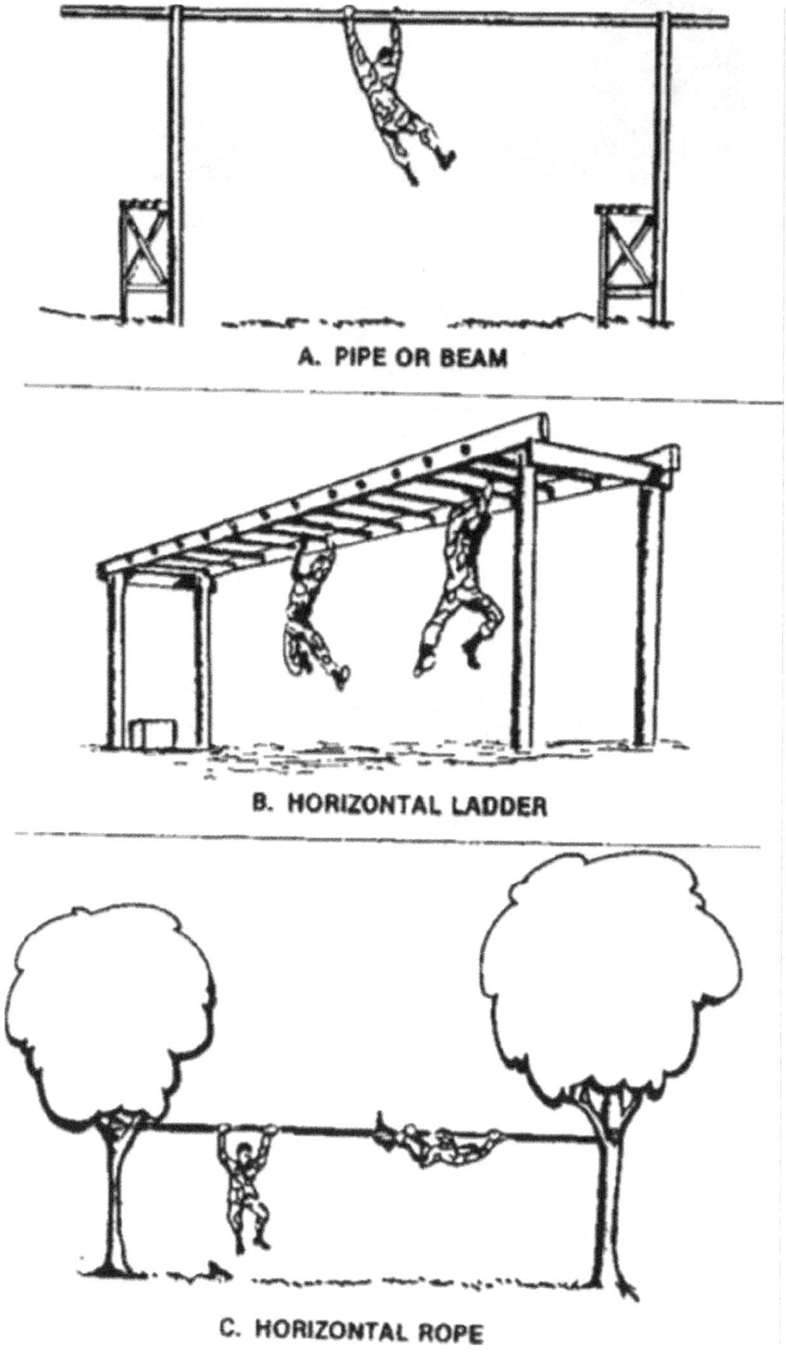

Figure 3-32.Horizontal Type Obstacles.

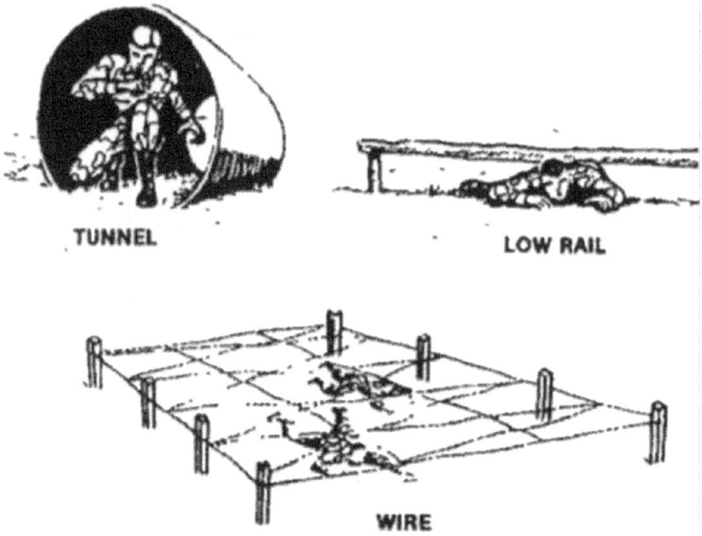

Figure 3-33.Crawling-Type Obstacles.

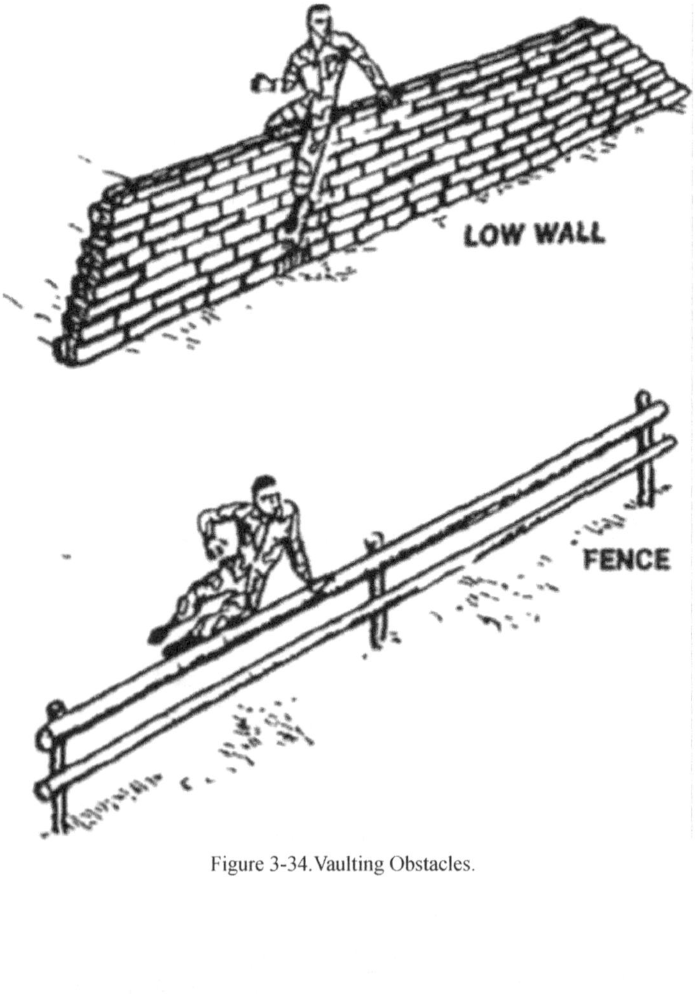

Figure 3-34.Vaulting Obstacles.

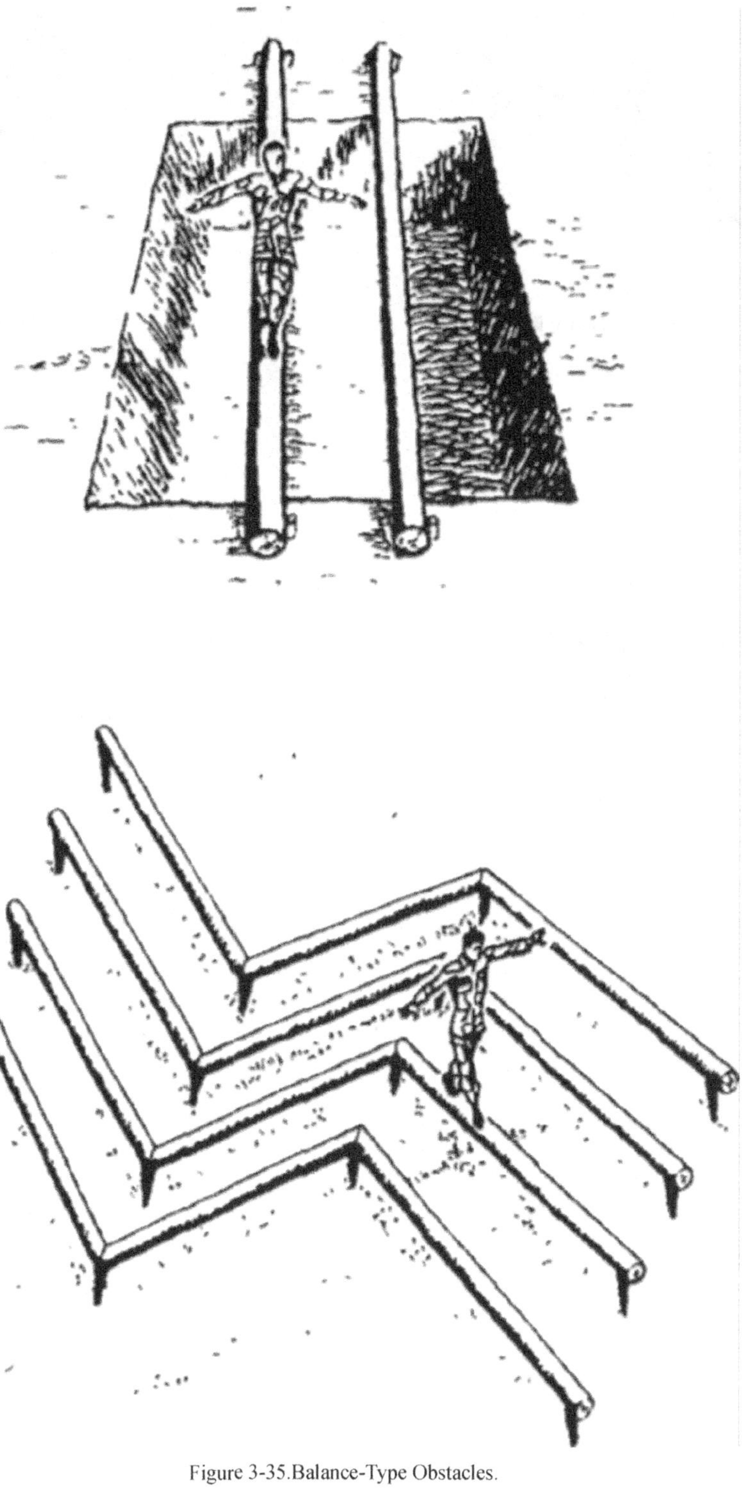

Figure 3-35.Balance-Type Obstacles.

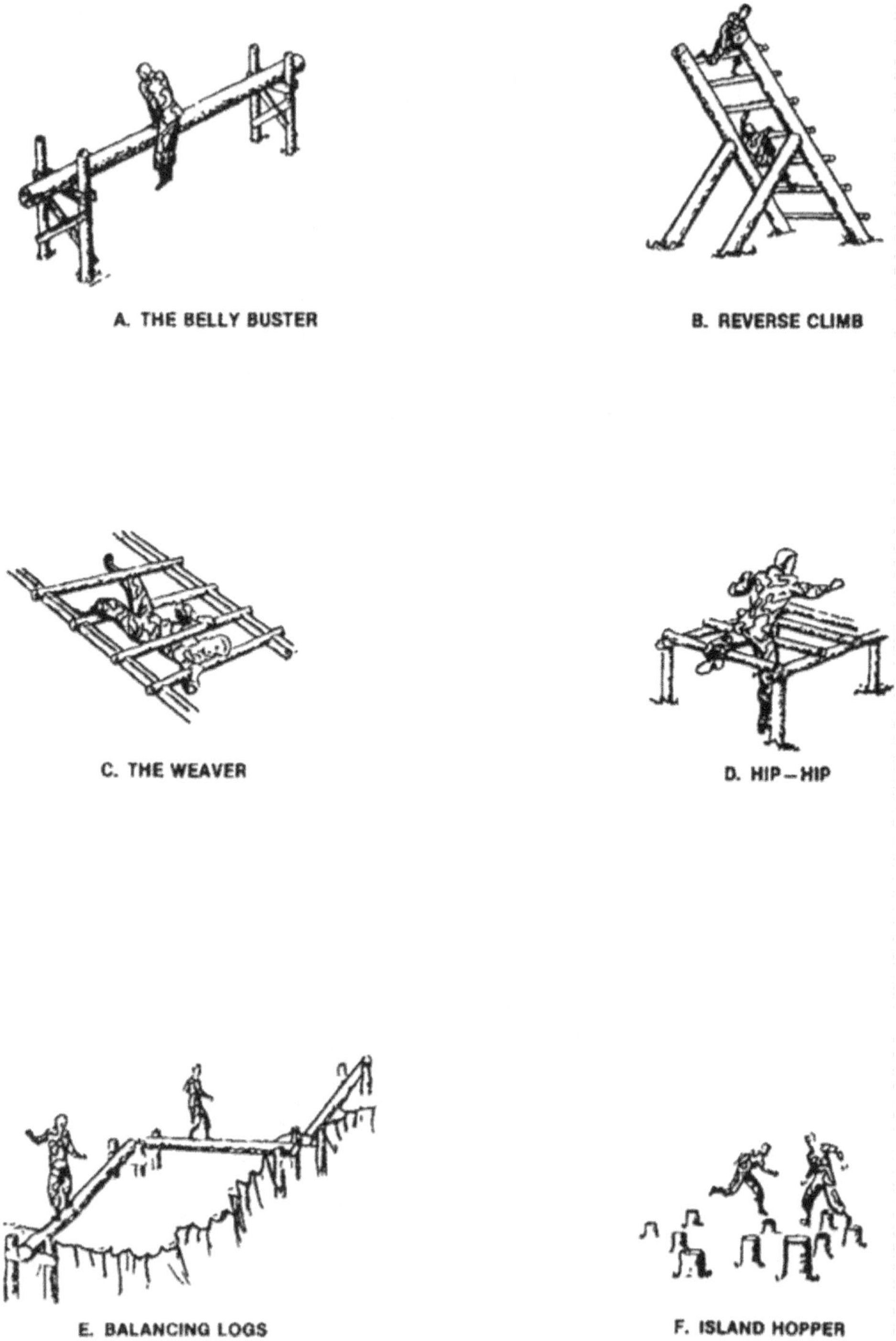

Figure 3-36.Red Group.

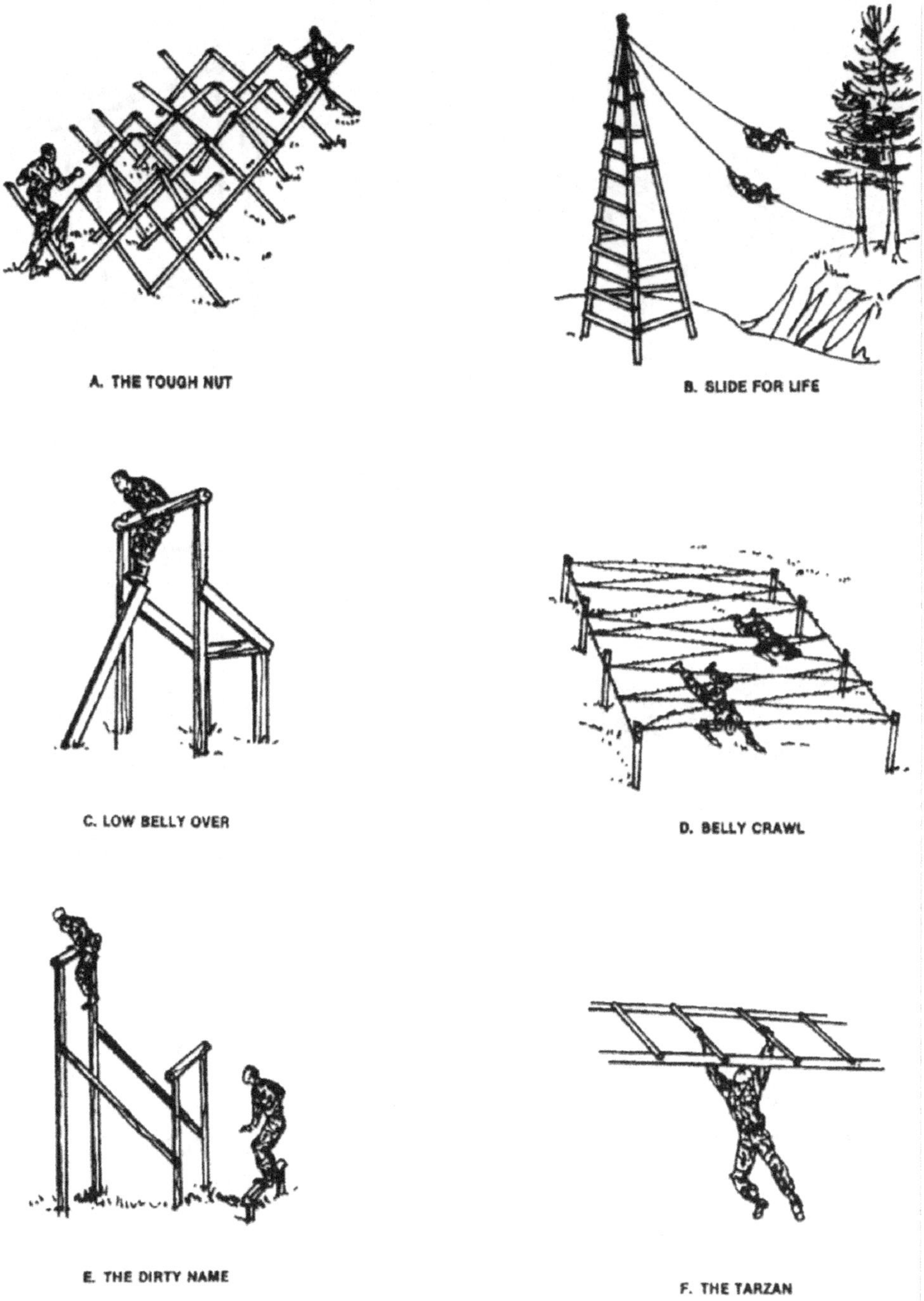

Figure 3-37.White Group.

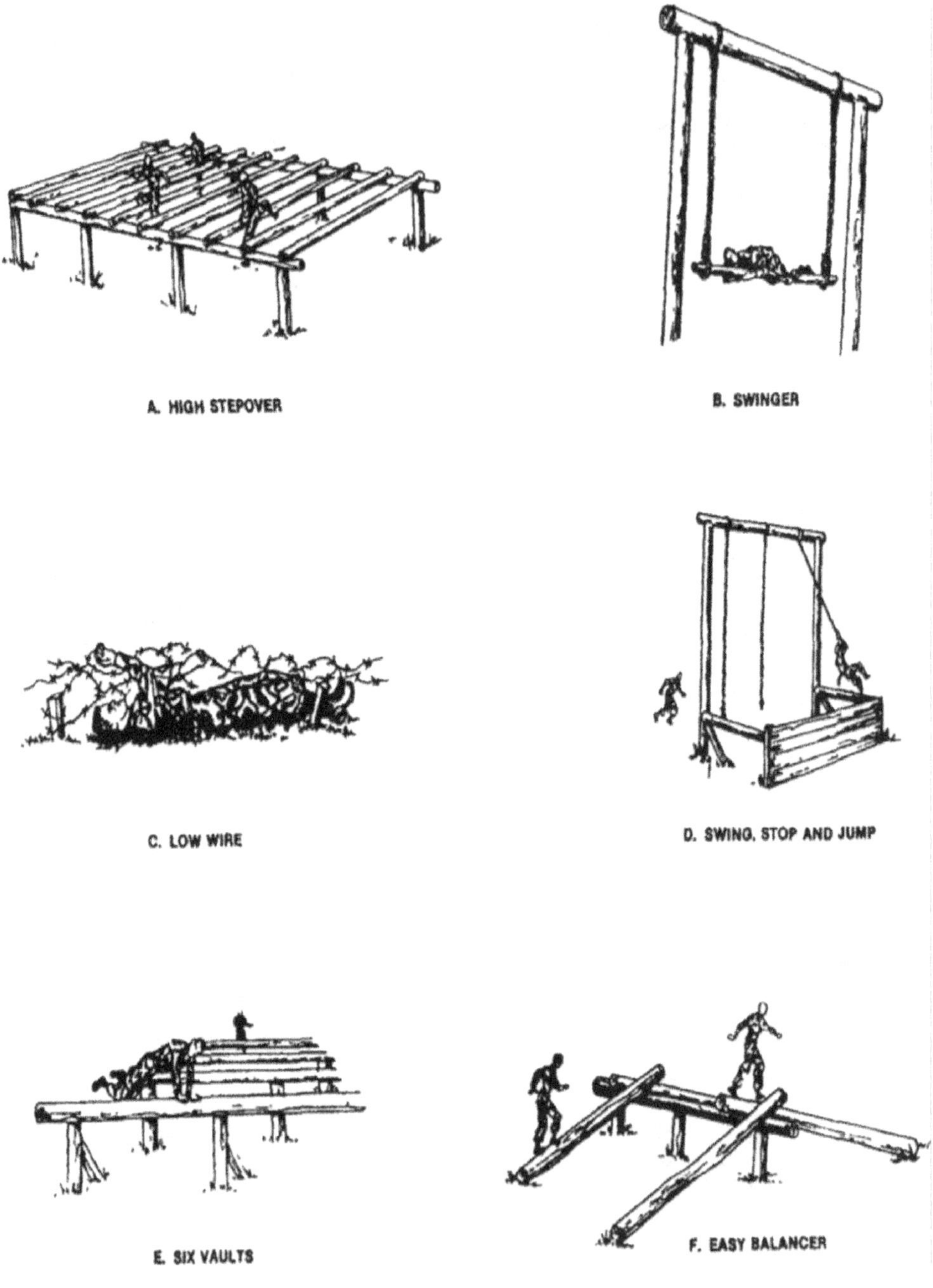

Figure 3-38.Blue Group.

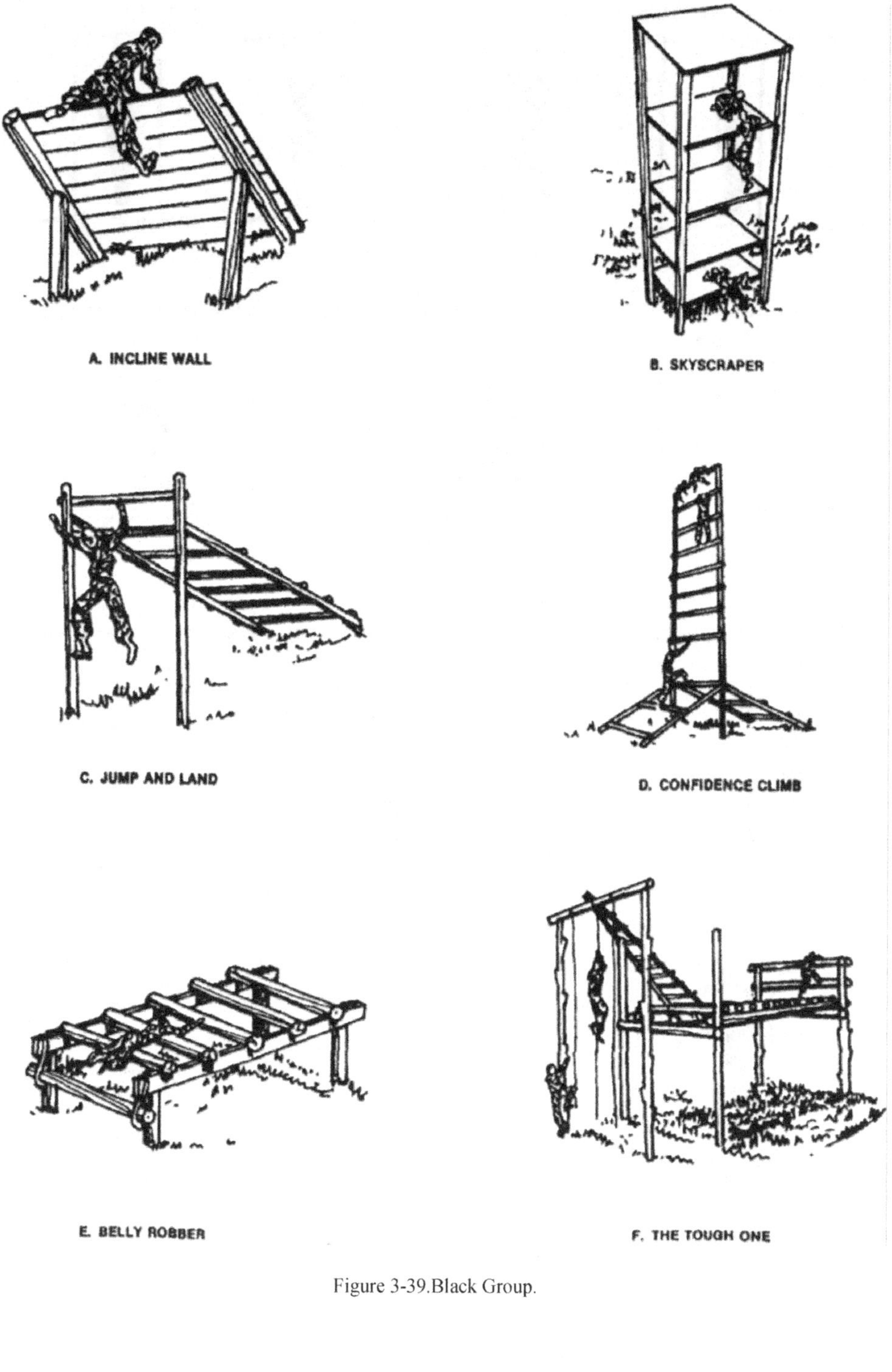

Figure 3-39.Black Group.

Figure 3-40. Bench Conditioning, Table I.

Figure 3-41. Bench Conditioning, Table II.

PROGRESSION GUIDE

AGE GROUP	LEVEL	EXERCISES 1	2	3	4	5	6
17	A	15	18	14	15	15	250
to	B	13	16	13	13	13	235
29	C	11	14	12	11	11	[illegible]
30	A	13	14	12	13	13	200
to	B	11	13	11	11	11	185
39	C	9	12	10	9	9	165
40	A	11	11	10	11	11	150
to	B	9	10	9	9	9	135
44	C	7	9	8	7	7	120
45	A	9	8	8	9	9	100
to	B	7	7	7	7	7	90
49	C	5	6	6	5	5	80
50	A	7	6	6	7	7	75
to	B	5	5	5	5	5	70
59	C	3	4	4	3	3	60
60	A	4	5	4	4	4	50
and	B	3	4	3	3	3	40
over	C	2	3	2	2	2	30
Minutes for each exercise		2	1	1	1	2	5

Table I. Progression Guide.

Table I. Exercises.

PROGRESSION GUIDE

AGE GROUP	LEVEL	EXERCISES 1	2	3	4	5	6
17	A	17	17	17	9	19	300
to	B	15	15	15	8	17	270
29	C	13	13	13	7	15	245
30	A	15	15	15	8	17	235
to	B	13	13	13	7	15	210
39	C	11	11	11	6	13	190
40	A	13	13	13	7	15	175
to	B	11	11	11	6	13	155
44	C	9	10	9	5	11	135
45	A	11	11	11	6	13	125
to	B	9	9	9	5	11	110
49	C	7	7	7	4	9	100
50	A	9	9	9	5	11	95
to	B	7	7	7	4	9	85
59	C	5	5	5	3	7	75
60	A	6	7	7	4	9	70
and	B	5	5	5	3	7	60
over	C	4	4	4	2	5	50
Minutes for each exercise		1	1	1	1 1/2	1 1/2	6

Table II. Progression Guide.

Table II. Exercises.

PROGRESSION GUIDE

AGE GROUP	LEVEL	EXERCISES 1	2	3	4	5	6
17	A	10	19	19	16	10	350
to	B	9	17	17	15	9	315
29	C	8	15	15	14	8	280
30	A	9	17	17	14	9	270
to	B	8	15	15	13	8	240
39	C	7	13	13	12	7	210
40	A	8	15	15	12	8	200
to	B	7	13	13	11	7	180
44	C	6	11	11	10	6	160
45	A	7	13	13	10	7	150
to	B	6	11	11	9	6	135
49	C	5	9	9	8	5	120
50	A	6	11	11	8	6	115
to	B	5	9	9	7	5	105
59	C	4	7	7	6	4	95
60	A	5	9	9	7	5	90
and	B	4	7	7	6	4	80
over	C	3	5	5	4	3	70
Minutes for each exercise		1 1/2	1	1	1 1/2	1	6

Table III. Progression Guide.

Table III. Exercises.

PROGRESSION GUIDE

AGE GROUP	LEVEL	EXERCISES 1	2	3	4	5	6
17	A	12	9	12	24	25	400
to	B	11	8	11	22	23	380
29	C	10	7	10	21	21	360
30	A	11	8	11	23	23	305
to	B	10	7	10	21	21	290
39	C	9	6	9	20	20	275
40	A	10	7	10	20	21	225
to	B	9	6	9	18	18	215
44	C	8	5	8	16	14	205
45	A	8	6	8	16	14	175
to	B	7	5	7	14	14	165
49	C	6	4	6	12	12	155
50	A	6	5	6	13	13	135
to	B	5	4	5	11	11	130
59	C	4	3	4	10	10	120
60	A	5	4	5	10	10	100
and	B	4	3	4	9	9	95
over	C	3	2	3	8	8	90
Minutes for each exercise		1	2	1	1	1	6

Table IV. Progression Guide.

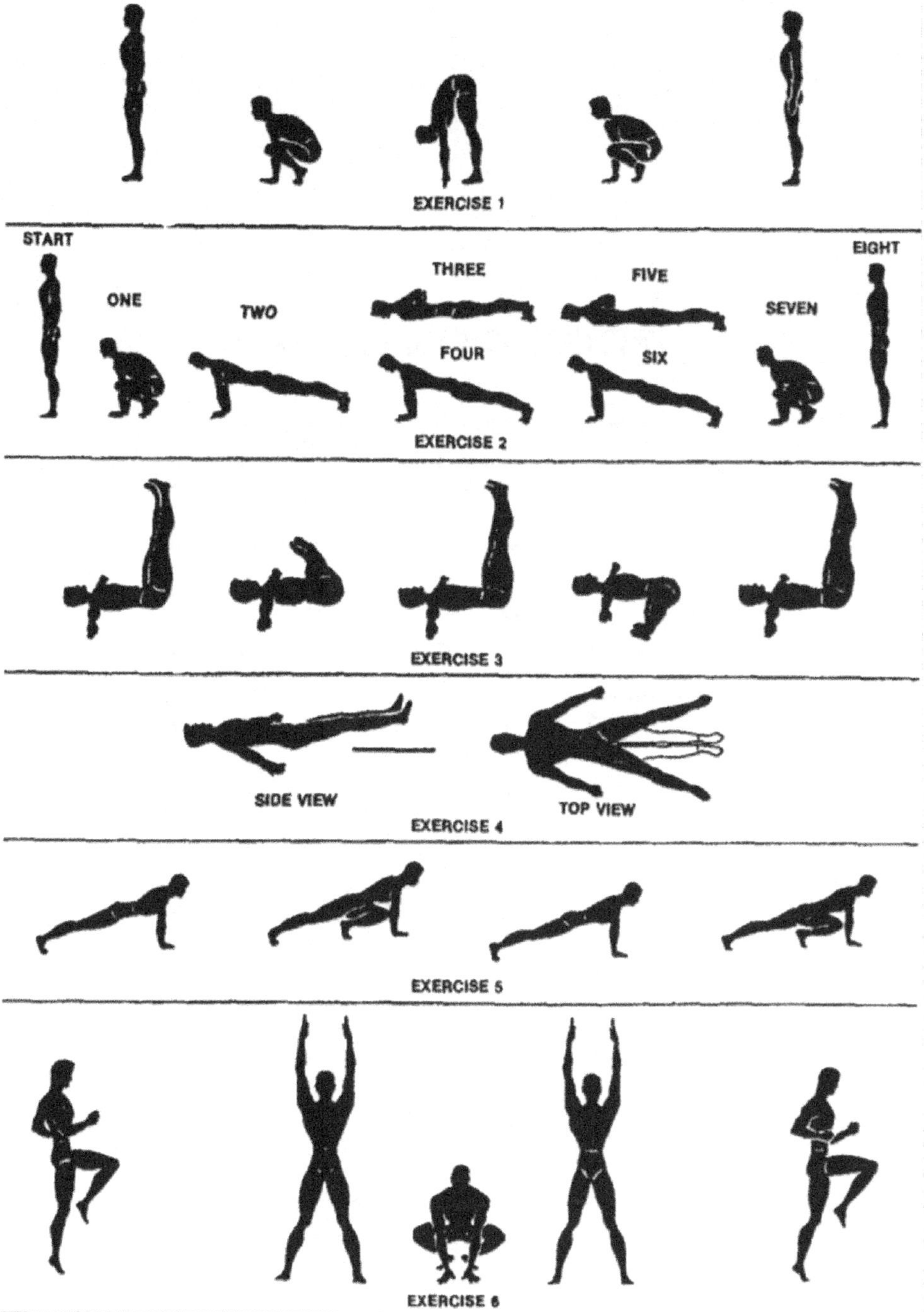

Table IV. Exercises.

PROGRESSION GUIDE

AGE GROUP	LEVEL	EXERCISES 1	2	3	4	5	6
17	A	14	13	28	14	30	450
to	B	13	12	27	13	28	430
29	C	12	11	26	12	26	410
30	A	12	12	25	12	26	350
to	B	11	11	24	11	24	330
39	C	10	10	23	10	22	310
40	A	11	11	23	11	23	250
to	B	10	10	21	10	21	240
44	C	9	9	19	9	19	230
45	A	9	9	20	9	20	200
to	B	8	8	18	8	18	190
49	C	7	7	16	7	16	180
50	A	7	7	16	7	16	170
to	B	6	6	14	6	14	155
59	C	5	5	12	5	12	140
60	A	6	6	12	6	12	115
and	B	5	5	11	5	10	110
over	C	4	4	9	4	9	105
Minutes for each exercise		2	1	1	2	1	5

Table V. Progression Guide.

Table V. Exercises.

PROGRESSION GUIDE

AGE GROUP	LEVEL	EXERCISES 1	2	3	4	5	6
17	A	17	15	32	32	35	500
to	B	16	14	30	30	33	480
29	C	15	13	28	28	31	460
30	A	15	13	30	30	31	400
to	B	14	12	28	28	29	380
39	C	13	11	26	26	27	360
40	A	13	10	27	27	27	310
to	B	12	9	25	25	25	285
44	C	11	8	23	23	23	265
45	A	11	9	23	23	23	250
to	B	10	8	21	21	21	230
49	C	9	7	19	19	19	210
50	A	9	8	19	19	19	200
to	B	8	7	17	17	17	190
59	C	7	6	15	15	15	175
60	A	8	7	15	15	17	140
and	B	7	6	13	13	15	130
over	C	5	5	10	10	12	120
Minutes for each exercise		2	1	1	1	1	6

Table VI. Progression Guide.

Table VI. Exercises.

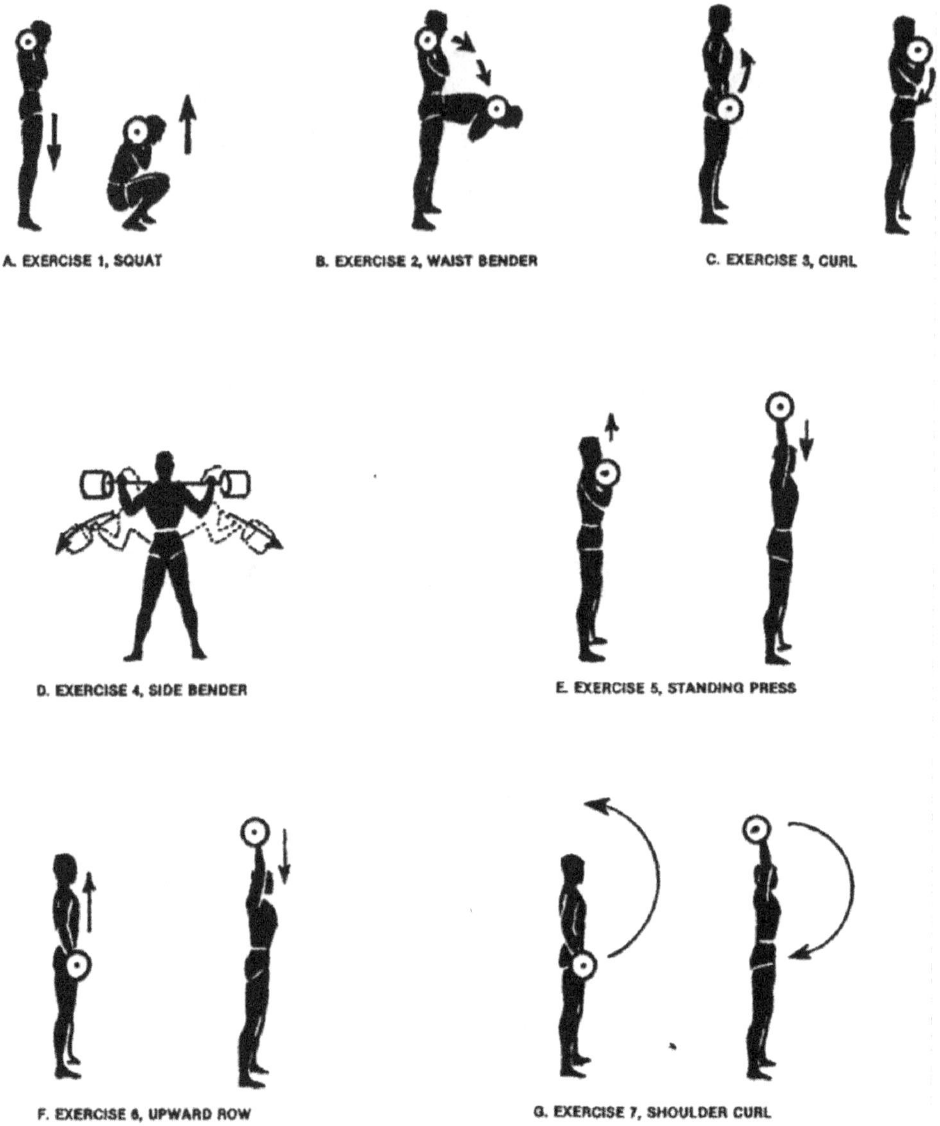

Figure 3-42. Weight Training.

A. EXERCISE 1, ARM PRESS

B. EXERCISE 2, LEG PRESS

C. EXERCISE 3, SIDE PRESS

D. EXERCISE 4, LATERAL RAISE

E. EXERCISE 5, NECK PRESS

F. EXERCISE 6, DOOR PULL

Figure 3-43. Isometric Exercises.

Figure 5-1.Combatives Table I.

Figure 5-2.Combatives Table II.

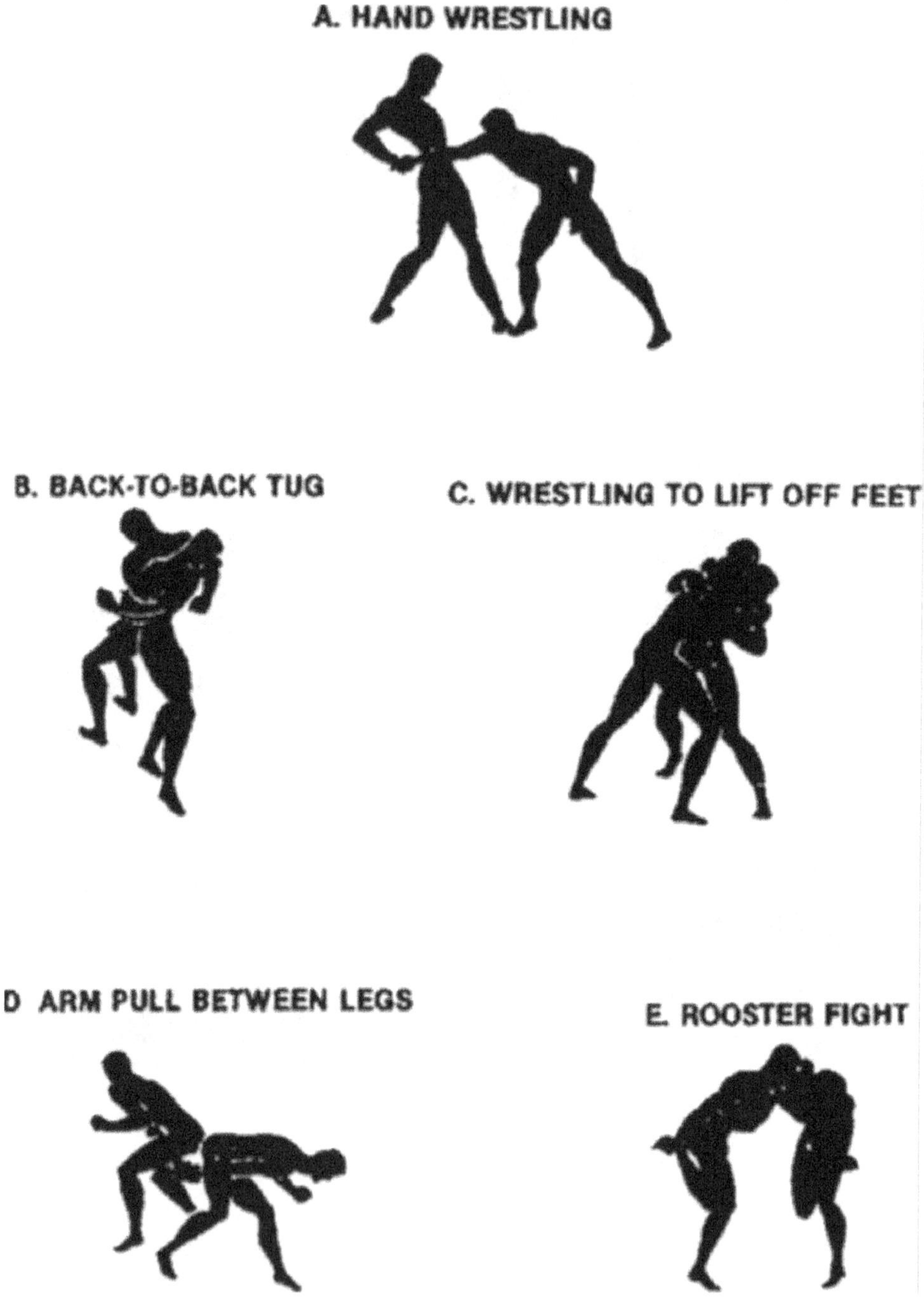

Figure. 5-3.Combatives Table III.

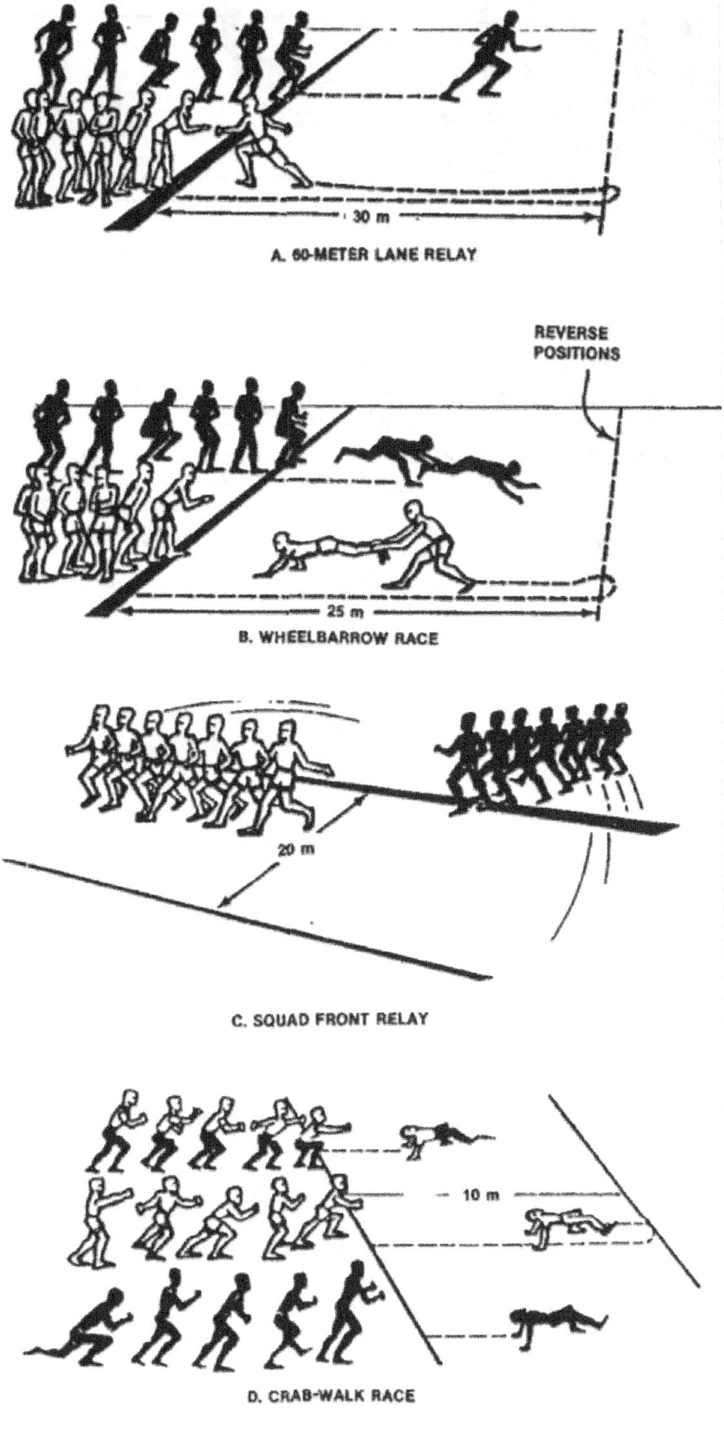

Figure 5-4.Relay Table I.

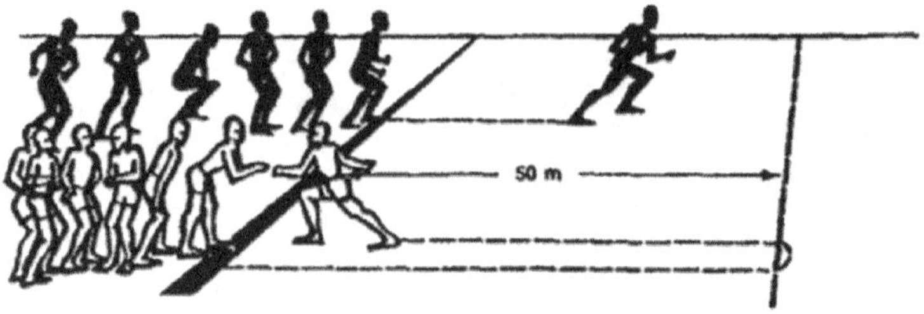

A. 100-METER LANE RELAY

B. FROG-JUMP RELAY

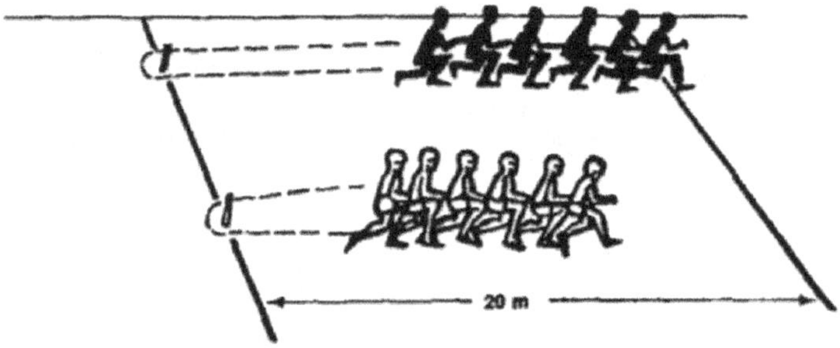

C. SIMPLE RELAY

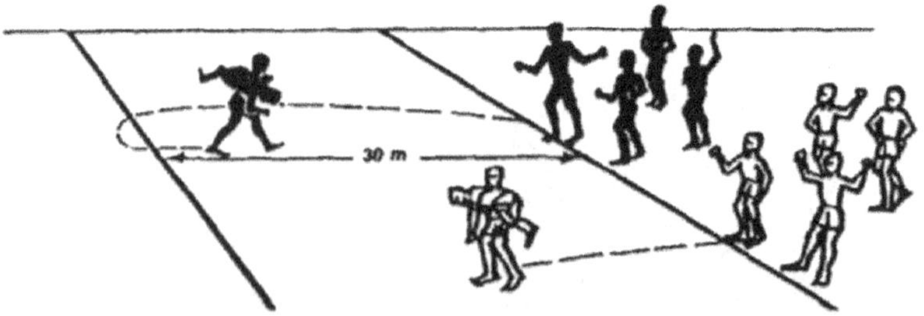

D. FIREMAN'S CARRY RELAY

Figure 5-5.Relay Table II.

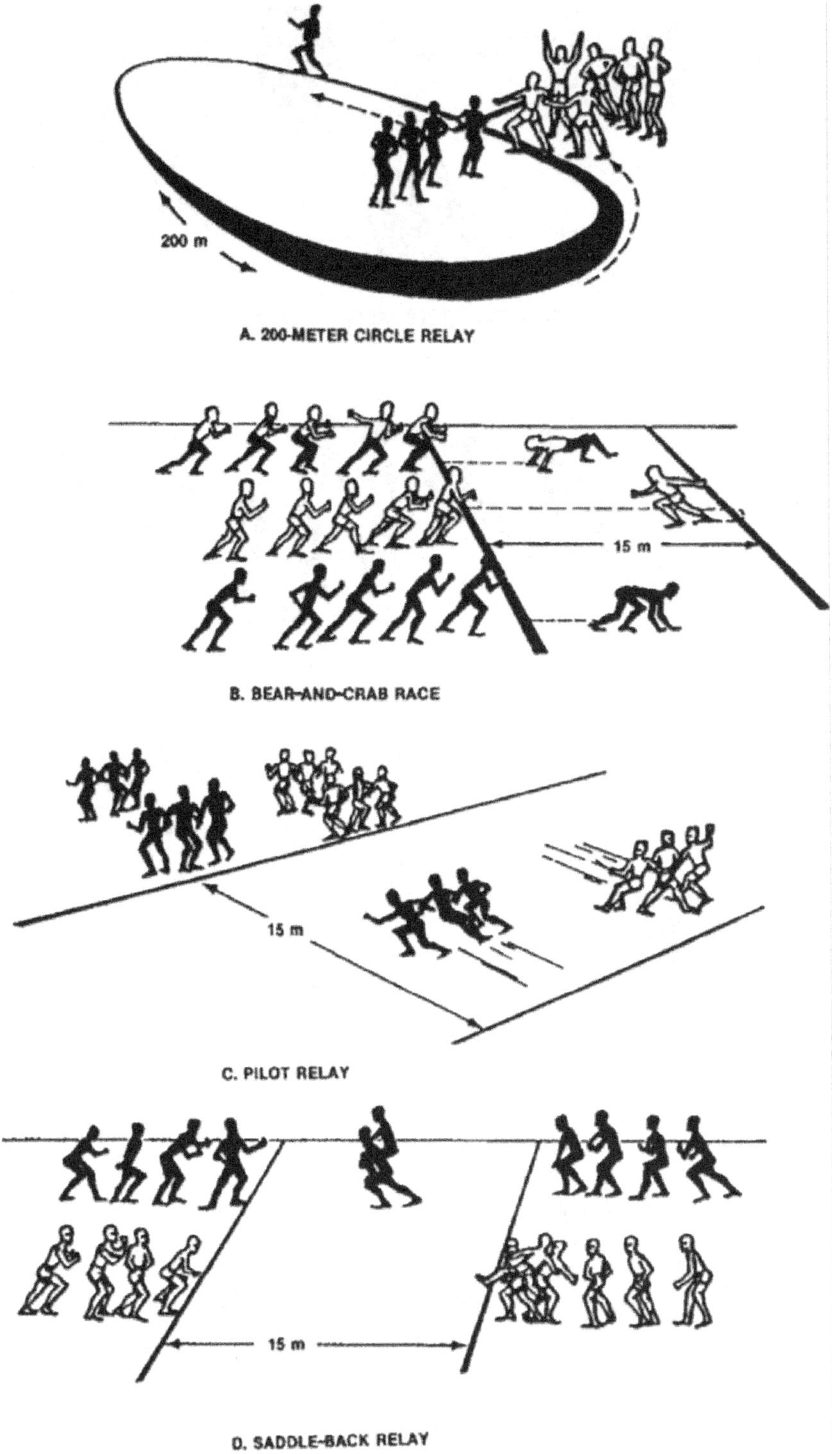

Figure 5-6.Relay Table III.

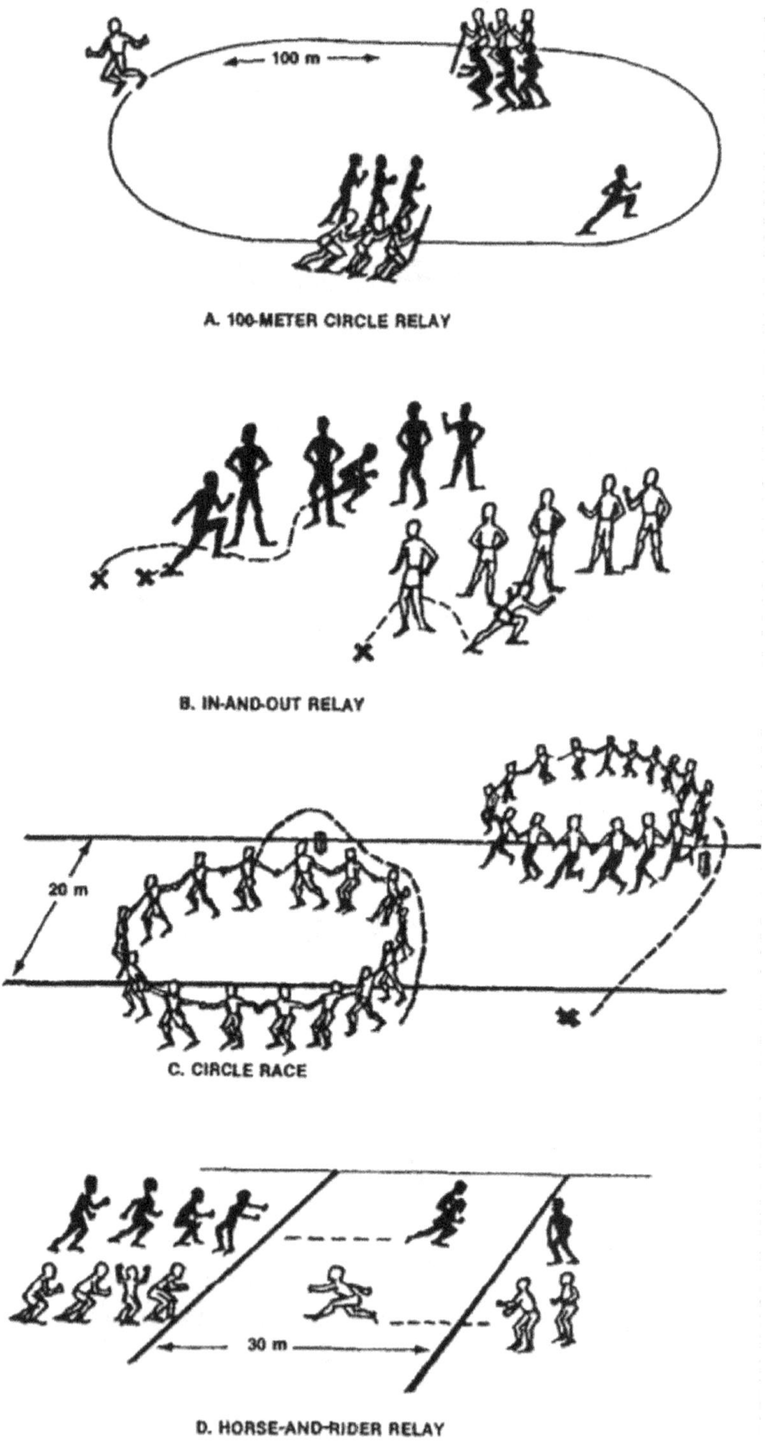

Figure 5-7.Relay Table IV.

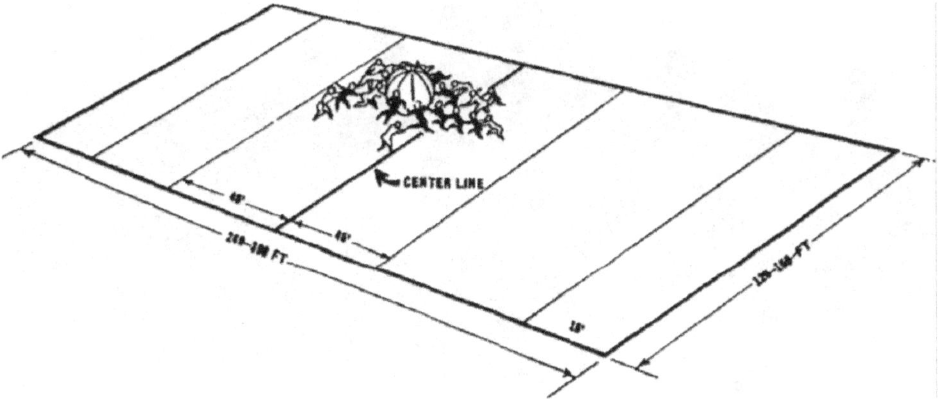

Figure 5-8.Pushball.

Figure 5-9.Line Rush.

Figure. 5-10. Human Tug-of-War.

Figure 5-11. Master of the Ring.

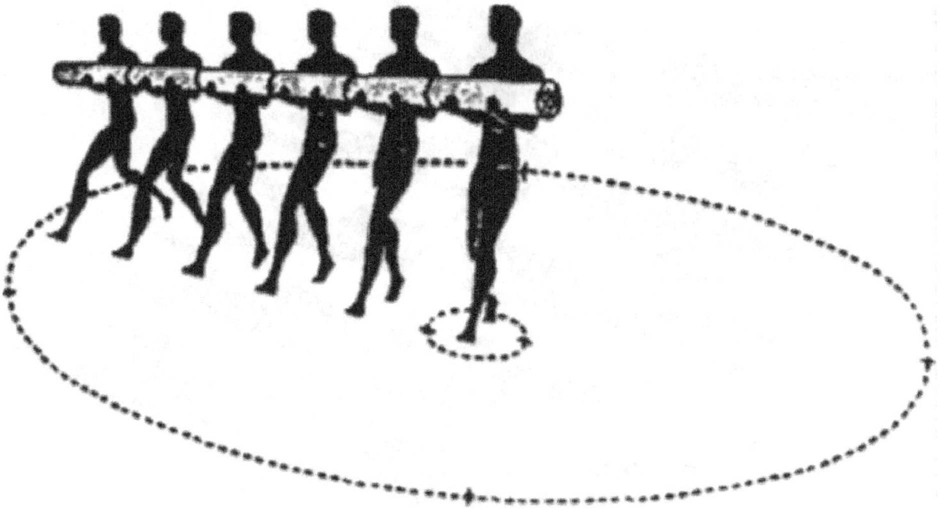

Figure 5-12.Log Pivot Circle.

Figure 5-13.Log-Rolling Race.

Figure 5-14. Prone Pushing Contest.

Figure 5-15. Shuttle Relay Race.

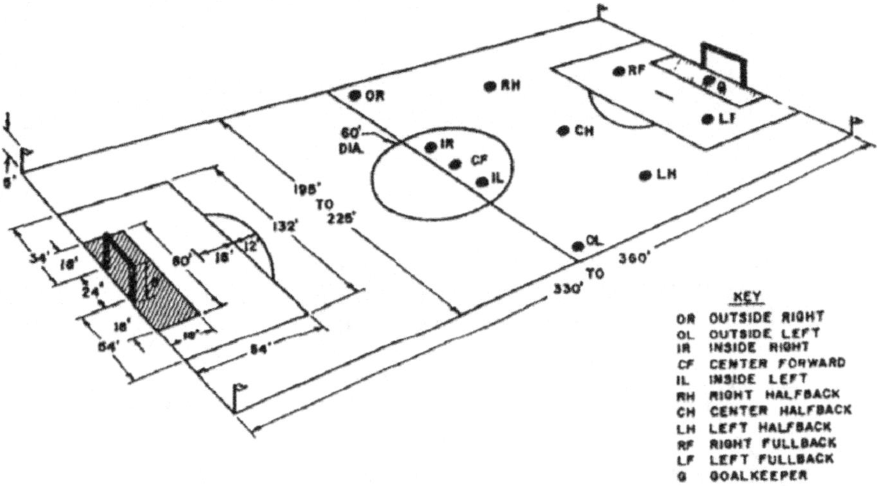

Figure 5-16. Soccer.

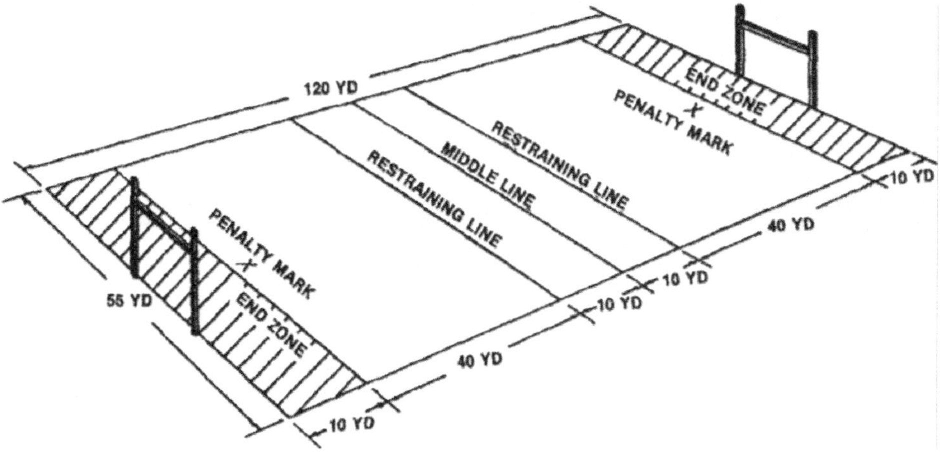

Figure 5-17.Speedball Field.

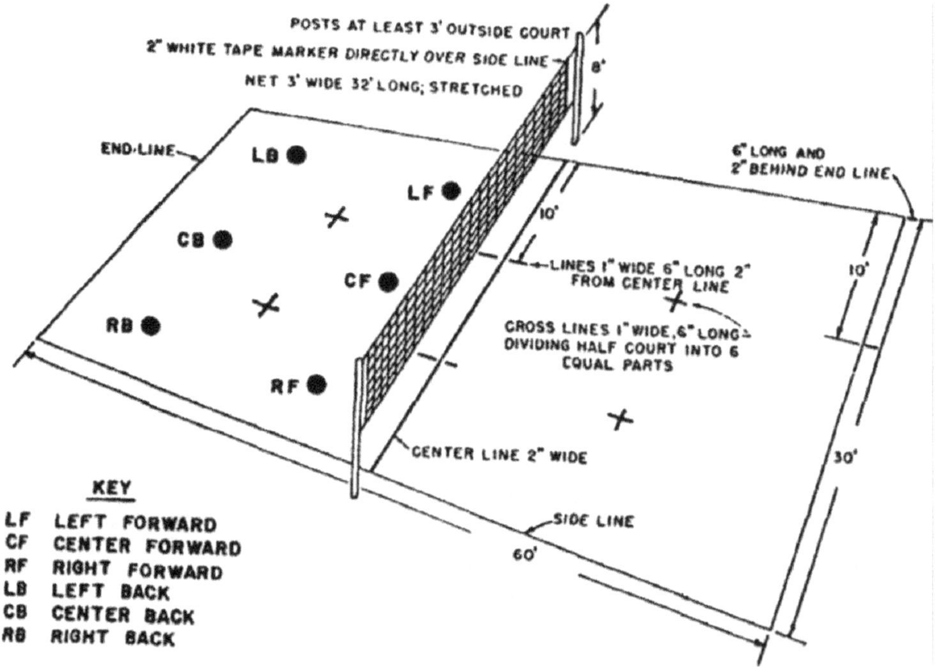

Figure 5-18.Volleyball.

PTS	UPHILL CLIMB	ROPE CLIMB	EVACUATION	FIRE AND MANEUVER	FORCED MARCH
100	1:20-BELOW	0:10-BELOW	0:25-BELOW	1:20-BELOW	26:00-BELOW
99	1:21-1:22	0:11		1:21-1:22	26:01-26:12
98	1:23-1:24	0:12	0:26	1:23-1:24	26:13-26:24
97	1:25-1:26	0:13		1:25-1:26	26:25-26:36
96	1:27-1:28	0:14	0:27	1:27-1:28	26:37-26:48
95	1:29-1:30	0:15		1:29-1:30	26:49-27:00
94	1:31-1:32	0:16	0:28	1:31-1:32	27:01-27:12
93	1:33-1:34	0:17		1:33-1:34	27:13-27:24
92	1:35-1:36	0:18	0:29	1:35-1:36	27:25-27:36
91	1:37-1:38	0:19		1:37-1:38	27:37-27:48
90	1:39-1:40	0:20	0:30	1:39-1:40	27:49-28:00
89	1:41-1:42	0:21	0:31	1:41-1:42	28:01-28:12
88	1:43-1:44	0:22	0:32	1:43-1:44	28:13-28:24
87	1:45-1:46	0:23	0:33	1:45-1:46	28:25-28:36
86	1:47-1:48	0:24	0:34	1:47-1:48	28:37-28:48
85	1:49-1:50	0:25	0:35	1:49-1:50	28:49-29:00
84	1:51-1:52	0:26	0:36	1:51-1:52	29:01-29:12
83	1:53-1:54	0:27	0:37	1:53-1:54	29:13-29:24
82	1:55-1:56	0:28	0:38	1:55-1:56	29:25-29:36
81	1:57-1:58	0:29	0:39	1:58-1:58	29:37-29:48
80	1:59-2:00	0:30	0:40	1:59-2:00	29:49-30:00
79	2:01-2:02	0:31	0:41	2:01-2:02	30:01-30:12
78	2:03-2:04	0:32	0:42	2:03-2:04	30:13-30:24
77	2:05-2:06	0:33	0:43	2:05-2:06	30:25-30:36
76	2:07-2:08	0:34	0:44	2:07-2:08	30:37-30:48
75	2:09-2:10	0:35	0:45	2:09-2:10	30:49-31:00
74	2:11-2:12	0:36	0:46	2:11-2:12	31:01-31:12
73	2:13-2:14	0:37	0:47	2:13-2:14	31:13-31:24
72	2:15-2:16	0:38	0:48	2:15-2:16	31:25-31:36
71	2:17-2:18	0:39	0:49	2:17-2:18	31:37-31:48
70	2:19-2:20	0:40	0:50	2:19-2:20	31:49-32:00
69	2:21-2:22	0:41	0:51	2:21-2:22	32:01-32:12
68	2:23-2:24	0:42	0:52	2:23-2:24	32:13-32:24
67	2:25-2:26	0:43	0:53	2:25-2:26	32:25-32:36
66	2:27-2:28	0:44	0:54	2:27-2:28	32:37-32:48
65	2:29-2:30	0:45	0:55	2:29-2:30	32:49-33:00

Figure 6-1.Physical Readiness Test Scoring Table for Male.

	EXAMPLE 1	EXAMPLE 2	EXAMPLE 3
Event #1	100 pts	100 pts	99 pts
Event #2	100 pts	99 pts	86 pts
Event #3	85 pts	75 pts (F)	80 pts
Event #4	90 pts	79 pts (F)	100 pts
Event #5	79 pts	85 pts	90 pts
Pts	454 pts	438 pts	455 pts
Final Score	79.0% (F)	69.0% (F)	91.0%
(F) = Failure			

Figure 6-2.Scoring Examples.

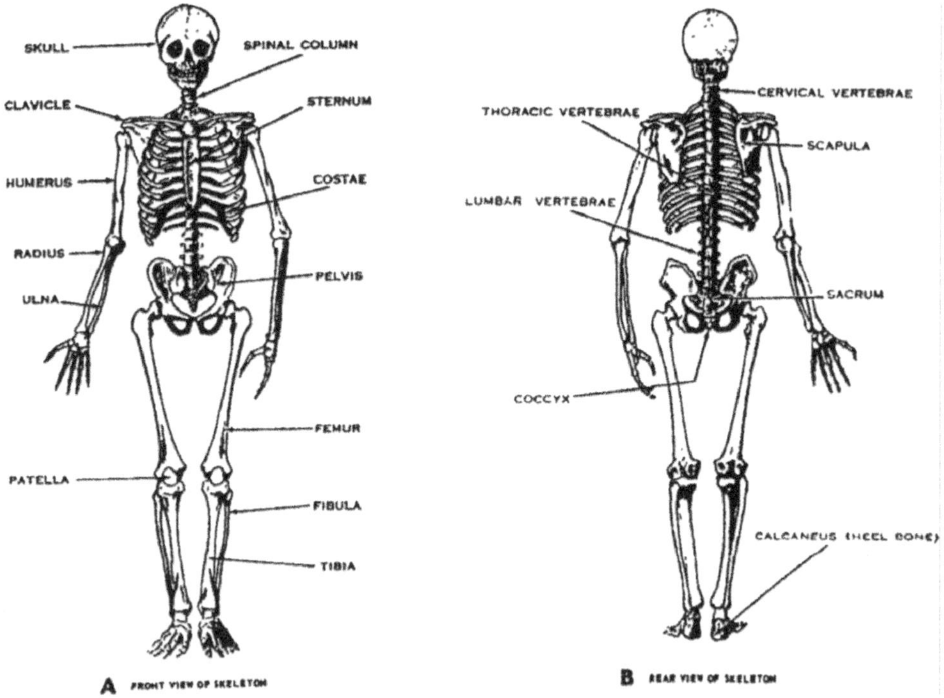

Figure 7-1.Front and Rear Views of Skeleton.

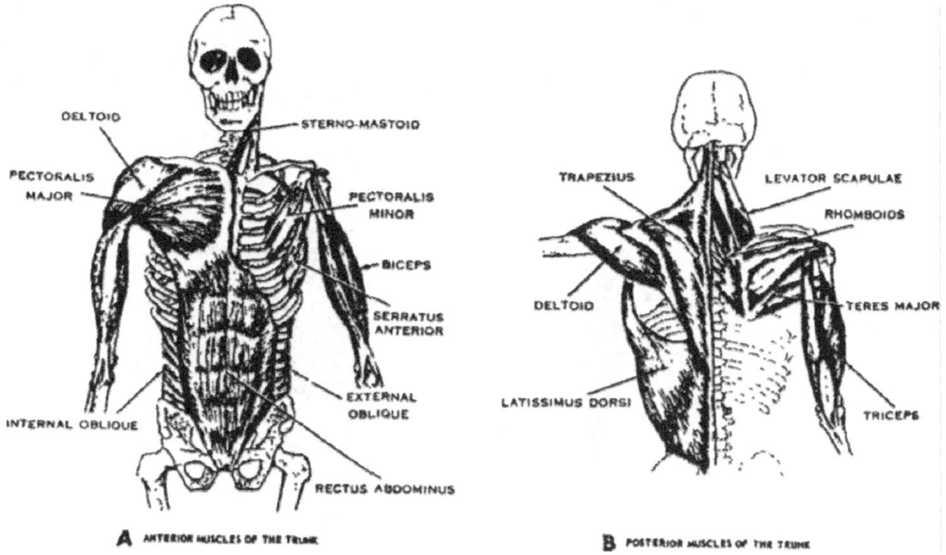

Figure 7-2.Muscles of the Trunk.

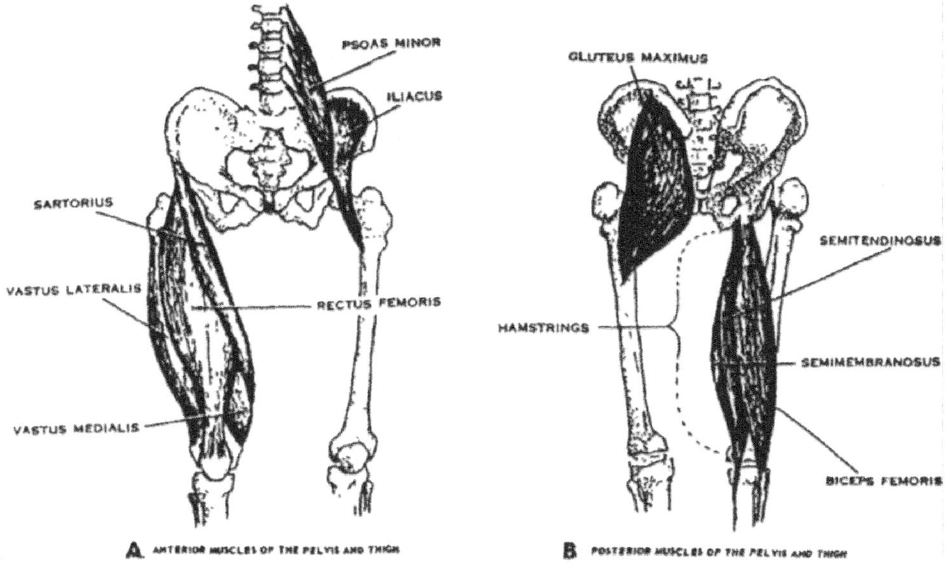

Figure 7-3.Anterior and Posterior Muscles of the Pelvisand Thigh.

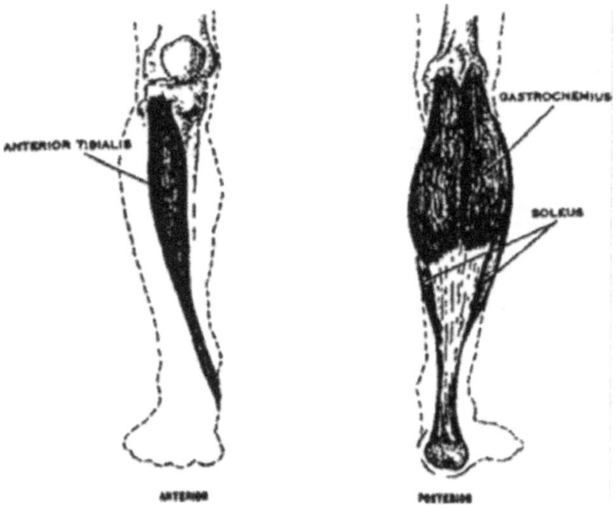

Figure 7-4. Muscles of Lower Leg.

Figure 7-5.Circulation of Blood from Heart to the Body.

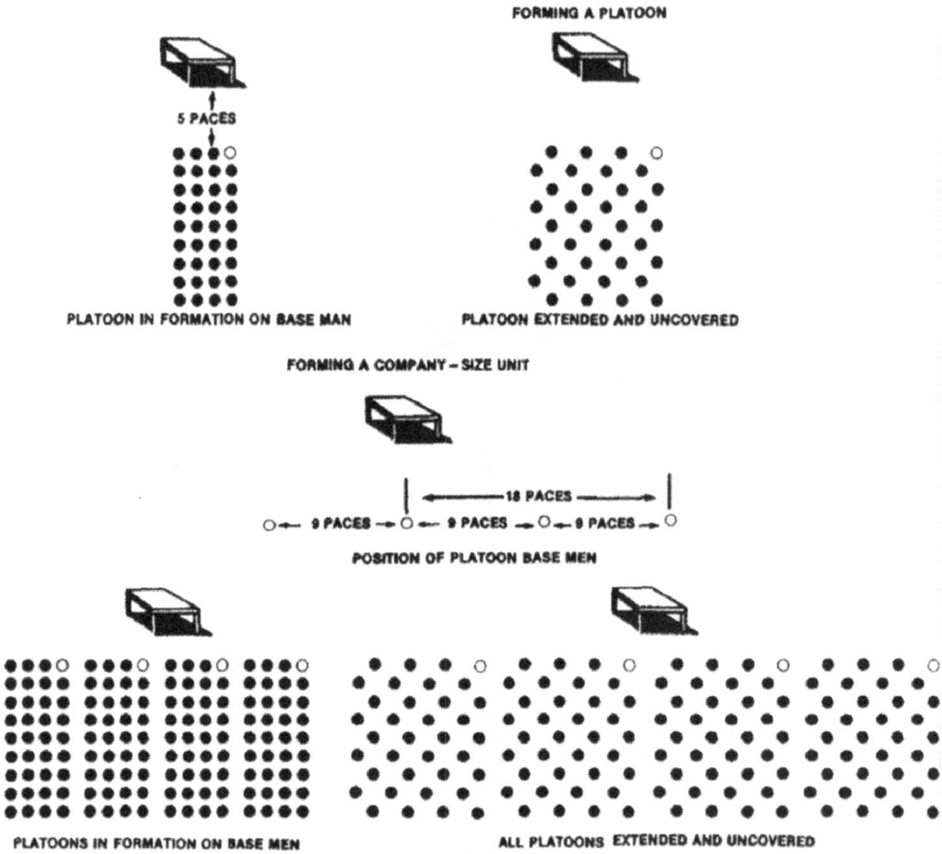

Figure A-l. Forming the Extended Rectangular Formation.

Figure A-2.Circle Formation.

DEPARTMENT OF THE NAVY
Headquarters United States Marine Corps
Washington, D.C. 20380-1775

18 February 1999

FOREWORD

1. PURPOSE

Today's Marines operate within a continuum of force where conflict may change from low intensity to high intensity over a matter of hours. Marines are also engaged in many military operations other than war, such as peacekeeping missions or noncombatant evacuation operations, where deadly force may not be authorized. During noncombative engagements, Marines must determine if a situation warrants applying deadly force. Sometimes Marines must decide in a matter of seconds because their lives or the lives of others depend on their actions. To make the right decision, Marines must understand both the lethal and nonlethal close combat techniques needed to handle the situation responsibly without escalating the violence unnecessarily. Marine Corps Reference Publication (MCRP) 3-02B, *Close Combat*, provides the tactics, techniques, and procedures of Marine Corps close combat. It also provides the doctrinal basis for the Marine Corps Close Combat Training Program (MCCCTP).

2. SCOPE

This publication guides individual Marines, unit leaders, and close combat instructors in the proper tactics, techniques, and procedures for close combat. MCRP 3-02B is not intended to replace supervision by appropriate unit leaders and close combat instruction by qualified instructors. Its role is to ensure standardization and execution of tactics, techniques, and procedures throughout the Marine Corps.

3. SUPERSESSION

MCRP 3-02B supersedes Fleet Marine Force Manual (FMFM) 0-7, *Close Combat*, dated 9 July 1993. There are significant differences between the two publications. MCRP 3-02B should be reviewed in its entirety.

4. WARNING

Techniques described in this publication can cause serious injury or death. Practical application in the training of these techniques will be conducted in strict accordance with approved Entry Level Close Combat, Close Combat Instructor (CCI), and Close Combat Instructor Trainer (CCIT) lesson plans. Where serious danger exists, the reader is alerted by the following:

WARNING

5. CERTIFICATION

Reviewed and approved this date.

BY DIRECTION OF THE COMMANDANT OF THE MARINE CORPS

J. E. RHODES
Lieutenant General, U.S. Marine Corps
Commanding General
Marine Corps Combat Development Command

DISTRIBUTION: 144 000066 00

OVERVIEW OF CLOSE COMBAT

1. Purpose of Close Combat

Close combat is the physical confrontation between two or more opponents. It involves armed and unarmed and lethal and nonlethal fighting techniques that range from enforced compliance to deadly force. The purpose of close combat is to execute armed and unarmed techniques to produce both lethal and nonlethal results. Unarmed techniques include hand-to-hand combat and defense against hand-held weapons. Armed techniques include techniques applied with a rifle, bayonet, knife, baton, or any weapon of opportunity.

2. Continuum of Force

Marines will find themselves in both combative and noncombative situations. The threat level in these situations can rise and fall several times based on the actions of both Marines and the people involved. The escalation of force stops when one person complies with the demands imposed by another person. This range of actions is known as a continuum of force. Continuum of force is the concept that there is a wide range of possible actions, ranging from voice commands to application of deadly force, that may be used to gain and maintain control of a potentially dangerous situation (MCO 5500.6_, *Arming of Security and Law Enforcement [LE] Personnel and the Use of Force*). Continuum of force consists of five levels that correspond to the behavior of the people involved and the actions Marines use to handle the situation (see the table below). Close combat techniques are executed in levels three, four, and five.

Level One: Compliant (Cooperative)

The subject complies with verbal commands. Close combat techniques do not apply.

Level Two: Resistant (Passive)

The subject resists verbal commands but complies immediately to any contact controls. Close combat techniques do not apply.

Continuum of Force		
Level	**Description**	**Actions**
1	Compliant (Cooperative)	Verbal commands
2	Resistant (Passive)	Contact controls
3	Resistant (Active)	Compliance techniques
4	Assaultive (Bodily Harm)	Defensive tactics
5	Assaultive (Serious Bodily Harm/ Death)	Deadly force
Note: Shading indicates levels in which Marines use close combat techniques.		

Level Three: Resistant (Active)

The subject initially demonstrates physical resistance. Marines use compliance techniques to control the situation. Level three incorporates close combat techniques to physically force a subject to comply. Techniques include—

- Come-along holds.
- Soft-handed stunning blows.
- Pain compliance through joint manipulation and the use of pressure points.

Level Four: Assaultive (Bodily Harm)

The subject may physically attack Marines, but he does not use a weapon. Marines use defensive tactics to neutralize the threat. Defensive tactics include the following close combat techniques:

- Blocks.
- Strikes.
- Kicks.
- Enhanced pain compliance procedures.
- Nightstick blocks and blows.

Level Five: Assaultive (Serious Bodily Harm/Death)

The subject usually has a weapon and will either kill or seriously injure someone if he is not stopped immediately and brought under control. Typically, to control the subject, Marines apply deadly force through the use of a firearm, but they may also use armed and unarmed close combat techniques.

3. Marine Corps Tactical Concepts

Close combat techniques support the following key Marine Corps tactical concepts. The concepts are not standalone ideas but are to be combined to achieve an effect that is greater than their separate sum.

Achieving a Decision

Achieving a decision is important in close combat. An indecisive fight wastes energy and possibly Marines' lives. Whether the intent is to control an opponent through restraint or defend themselves in war, Marines must have a clear purpose before engaging in close combat and act decisively once engaged.

Gaining an Advantage

A basic principle of martial arts is to use the opponent's strength and momentum against him to gain more leverage than one's own muscles alone can generate, thereby gaining an advantage. In close combat, Marines must exploit every advantage over an opponent to ensure a successful outcome. This can include employing various weapons and close combat techniques that will present a dilemma to an opponent. Achieving surprise can also greatly increase leverage. Marines try to achieve surprise through deception, stealth, and ambiguity.

Speed

Marines use speed to gain the initiative and advantage over the enemy. In close combat, the speed and violence of the attack against an opponent provides Marines with a distinct advantage. Marines must know and understand the basics of close combat so they can act instinctively with speed to execute close combat techniques.

Adapting

Close combat can be characterized by friction, uncertainty, disorder, and rapid change. Each situation is a unique combination of shifting factors that cannot be controlled with precision or certainty. For example, a crowd control mission may call for Marines to employ various techniques ranging from nonlethal restraint to more forceful applications. Marines who adapt quickly will have a significant advantage.

Exploiting Success

Typically, an enemy will not normally surrender simply because he was placed at a disadvantage. Marines cannot be satisfied with gaining an advantage in a close combat situation. They must exploit any advantage aggressively and ruthlessly until an opportunity arises to completely dominate the opponent. Marines must exploit success by using every advantage that can be gained.

(reverse blank)

Close Combat

Table of Contents

CHAPTER 1

FUNDAMENTALS OF CLOSE COMBAT

This chapter describes all techniques for a right-handed person. However, all techniques can be executed from either side.

The Marine is depicted in camouflage utilities. The opponent is depicted without camouflage.

The fundamentals of close combat include ranges, weapons of the body, target areas of the body, and pressure points of the body. These fundamentals form the basis for all close combat techniques. They provide Marines with a common framework regardless of the type of confrontation or the techniques used. If Marines apply these fundamentals properly in a close combat situation, they may save their lives or the lives of fellow Marines.

1. Ranges of Close Combat

Close combat engagements occur within three ranges: long range, midrange, and close range. During any engagement, these ranges may blur together or may rapidly transition from one to another until either the opponent is defeated or the conflict is resolved.

Long Range

During long range engagements, combatants engage each other with rifles, bayonets, sticks, or entrenching tools. See figure below.

Midrange

During midrange engagements, combatants engage each other with knives, punches, or kicks.

Close Range

During close range engagements, combatants grab each other. Close range engagements also involve elbow strikes, knee strikes, and grappling.

2. Weapons of the Body

Hands and Arms

The hands, forearms, and elbows are the arm's individual weapons. The hands consist of several areas that can be used as weapons: fists, edges of hands, palms, and fingers.

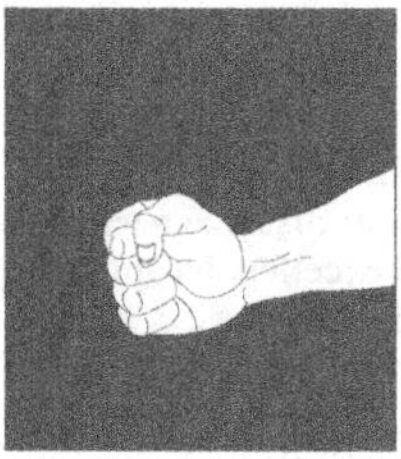

Fists. To minimize injury to the fists, Marines use their fists as weapons to target soft tissue areas such as the throat. The fists' striking surfaces are the first two knuckles of the hands or the meaty portions of the hands below the little fingers.

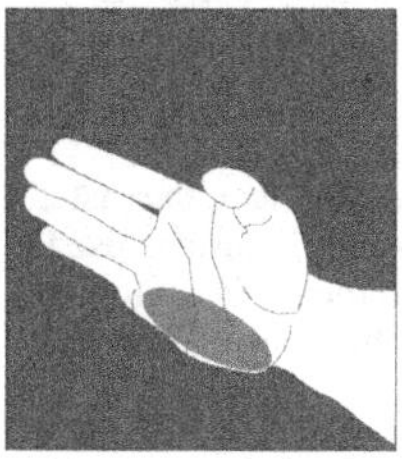

Edge of Hand. Marines use the edge of the hand (knife edge) as a weapon. Marines use the edge of the hand to strike soft tissue areas.

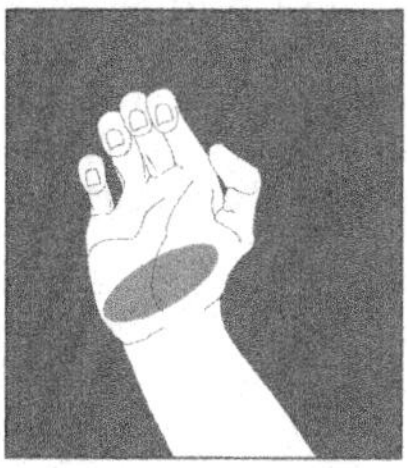

Palms. Because of the palm's padding, Marines use the heels of the palms to strike, parry, and/or block.

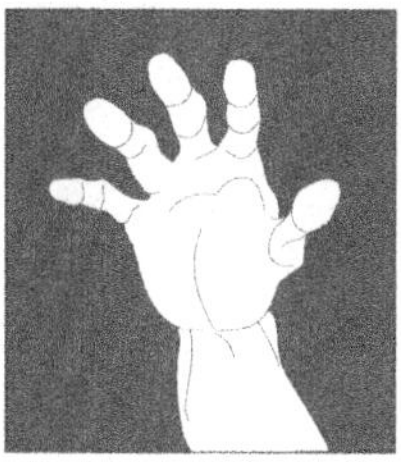

Fingers. Marines use the fingers to gouge, rip, and tear soft tissue areas (e.g., eyes, throat, groin).

Forearms. Marines use the forearms as a defensive tool to deflect or block attacks. Forearms can also be used as striking weapons to damage or break an opponent's joints and limbs. Marines sustain less self-injury when strikes are conducted with the forearms than when strikes are conducted with fists and fingers.

Elbows. Marines use the elbows as striking weapons. Because of the short distance needed to generate power, elbows are excellent weapons for striking during the close range of close combat.

Legs

The legs are more powerful than any other weapon of the body, and they are less prone to injury when striking. The feet are protected by boots and are the preferred choice for striking.

Feet. Marines use the balls of the feet, the insteps, and the toes to kick an opponent. Marines use the cutting edge of the heels and the heels to stomp on an opponent. Marines must be wearing boots when striking with the toes.

Knees. Like elbows, knees are excellent weapons in the close range of close combat. Knee strikes are most effective while fighting close to an opponent where kicks are impractical. The opponent's groin area is an ideal target for the knee strike if he is standing upright. Knee strikes can deliver a devastating secondary attack to an opponent's face following an initial attack that caused him to bend at the waist.

3. Target Areas of the Body

During close combat, Marines strive to attack the accessible target areas of an opponent's body. The readily accessible areas will vary with each situation and throughout the engagement. The target areas are divided into five major groups: head, neck, torso, groin, and extremities. The figure below illustrates target areas of the body.

Head

The vulnerable regions of the head are the eyes, temple, nose, ears, and jaw. Massive damage to the head kills an opponent.

Eyes. The eyes are excellent targets because they are soft tissue areas that are not protected by bone or muscle. Attacks to this area may cause the opponent to protect the area with his hands, allowing Marines to execute a secondary attack to other

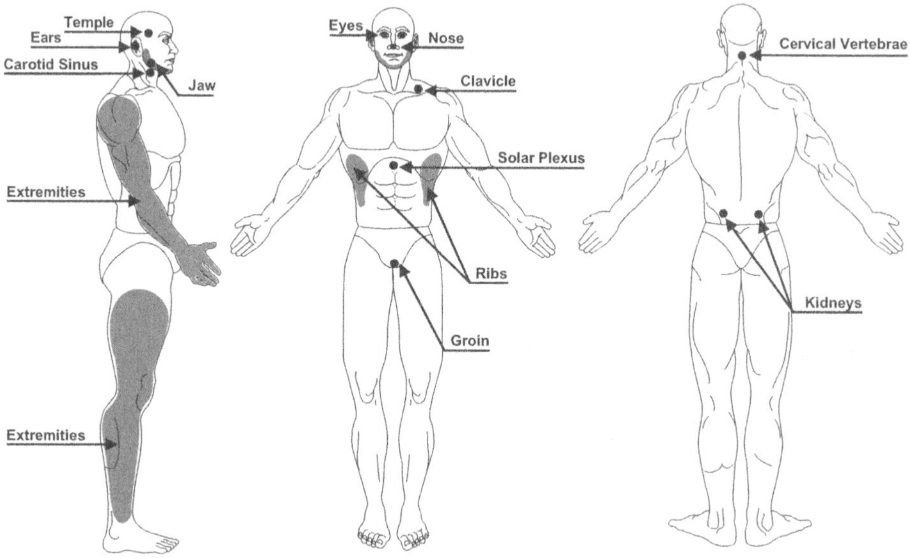

target areas while the opponent uses his hands to protect his eyes.

Temple. The temple is one of the most fragile areas of the head. Powerful strikes to the opponent's temple cause permanent damage and death.

Nose. The nose is very sensitive and easily broken. An attack to this area causes involuntary watering and closing of the opponent's eyes, rendering him vulnerable to secondary attacks. However, through training, individuals can condition themselves to withstand attacks to the nose. Therefore, any attack to the nose must be powerfully delivered.

Ears. Attacks to the ears may cause the eardrum to rupture. But this may not stop or even distract an opponent unless Marines powerfully deliver the strike.

Jaw. The jaw region, when struck forcefully, renders the opponent unconscious. Strikes to the jaw cause painful injuries to the teeth and surrounding tissues (e.g., lips, tongue), but the risk of self-injury is great unless Marines deliver strikes with a hard object such as a helmet, rifle butt, or boot heel.

Neck

The front of the neck, or throat area, is a soft tissue area that is not covered by natural protection. Damage to this region causes the opponent's trachea to swell, closing his airway, which can lead to death.

Carotid Sinus. The carotid sinus is located on both sides of the neck just below the jaw. Strikes to the carotid sinus restrict blood flow to the brain, causing loss of consciousness or death.

Cervical Vertebrae. The cervical vertebrae on the back of the neck, from the base of the skull to the top of the shoulders, contains the spinal cord, which is the nervous system's link to the brain. The weight of the head and the lack of large muscle mass allow damage to the cervical vertebrae and spinal cord. Excessive damage to this area causes pain, paralysis, or death.

Torso

Clavicle. The opponent's clavicle (or collar bone) can be easily fractured, causing immobilization of the arm.

Solar Plexus. Attacks to the opponent's solar plexus or center of the chest can knock the breath out of him and immobilize him.

Ribs. Damage to the opponent's ribs immobilizes him. It may also cause internal trauma.

Kidneys. Powerful attacks to the opponent's kidneys cause immobilization, permanent damage, or death.

Groin

The groin area is another soft tissue area not covered by natural protection. Any damage to this area causes the opponent to involuntarily protect his injured area, usually with his hands or legs. In male opponents, the scrotum is the main target since even a near miss causes severe pain, contraction of the lower abdominal muscles, deterioration of his stance, and possible internal trauma.

Extremities

Rarely will an attack to the opponent's extremities (arms and legs) cause death, but they are important target areas in close combat. Damage to an opponent's joints causes immobilization.

4. Pressure Points of the Body

There are nerves in the human body that, when pressure is applied or when they are struck, allow Marines to control a subject through pain compliance. Marines use pressure points to control an opponent when deadly force is not authorized. They also use pressure points to soften or distract an opponent so a lethal or nonlethal technique can be employed. The figure on page 1-5 illustrates the body's pressure points. Marines execute attacks to pressure points by—

- Rapidly kicking or striking pressure points.
- Slowly applying steady pressure to pressure points.

Infraorbital Nerve

The infraorbital nerve is just below the nose. Marines apply pressure to this nerve with an index finger to control the opponent.

Mastoid Process

The mastoid process is behind the base of the ear and beneath the edge of the jaw. Marines apply inward and upward pressure to this pressure point with the fingers to distract and control the opponent.

Jugular Notch

The jugular notch is at the base of the neck in the notch formed at the center of the clavicle. Marines apply pressure in a quick, stabbing motion with the index finger. Strikes to the jugular notch cause serious damage.

Brachial Plexus (Tie In)

The brachial plexus (tie in) is on the front of the shoulder at the joint. Strikes and pressure applied with the hand are effective on this nerve.

Radial Nerves

Radial nerves are on the inside of the forearms along the radius bones. Strikes and pressure applied with the hand to the radial nerve serve as a softening technique.

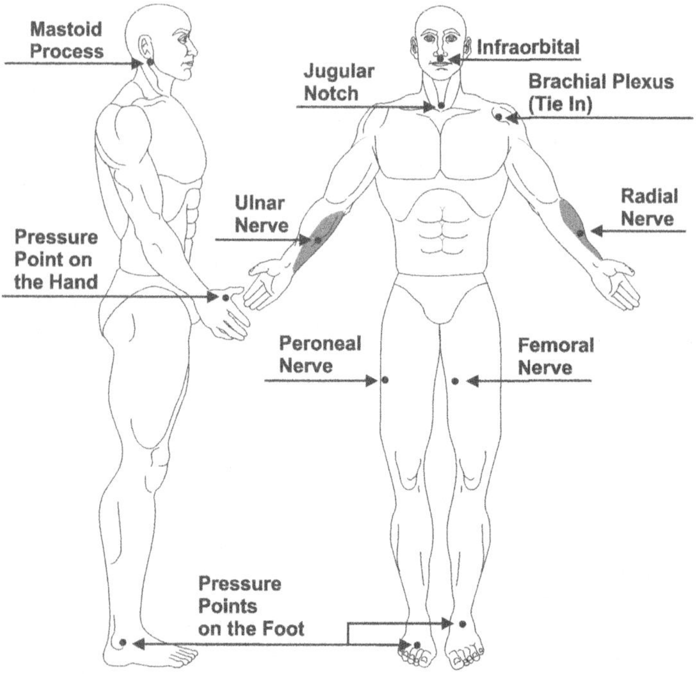

Ulnar Nerve

Ulnar nerves are on the outside of the forearms along the ulnar bones. Strikes and pressure applied with the hand to the ulnar nerve serve as a softening technique.

Pressure Point on the Hand

The hands contain a pressure point on the webbing between the thumbs and index fingers where the two bones of the fingers meet.

To force an opponent to soften or release his grip, Marines apply pressure with their index fingers to this pressure point or strike this pressure point with their fists.

Femoral Nerves

Femoral nerves are on the inside of the thighs along the femur bones. Strikes to the femoral nerve serve as a softening technique.

Peroneal Nerves

Peroneal nerves are on the outside of the thighs along the femur bones. Strikes to the peroneal nerve serve as a softening technique.

Pressure Points on the Feet

There are pressure points on the feet that, when pressure is applied or when they are struck, serve to soften or distract the opponent. Marines apply pressure with the toe, edge, or heel of their boots to the following points:

- The notch below the ball of the ankle.
- The top center of the foot, above the toes.
- The top of the foot where the leg and foot meet.

5. Basic Warrior Stance

Achieving a solid stance is critical to stability and movement throughout any close combat confrontation. The basic warrior stance provides the foundation for stability and movement that is needed to execute close combat techniques. To execute the basic warrior stance, Marines put their feet apart, hands up, elbows in, and chin down.

Feet Apart

Place feet shoulder-width apart.

Keep the head forward and eyes on the opponent, take a half step forward with the left foot, and pivot on the heels so the hips and shoulders are at approximately a 45-degree angle to the right.

Distribute body weight evenly on both legs. Bend the knees slightly.

Hands Up

Curl the fingers naturally into the palm of the hand. Position the thumb across the index and middle fingers. Do not clench the fists. Clenching the fists increases muscular tension in the forearms and decreases speed and reaction time.

Bring the hands up to the face at chin level, with the palms facing each other. Hold the fists up high enough to protect the head, but not so high that they block the field of vision. Ensure continuous eye contact with the opponent.

Elbows In

Tuck the elbows in close to protect the body.

Chin Down

Tuck the chin down to take advantage of the natural protection provided by the shoulders.

6. Angles of Approach and Movement

Marines use movement to control a confrontation and to retain a tactical advantage. Movement increases power and maximizes momentum. By moving around the opponent, Marines gain access to different target areas of the opponent's body. Movement allows Marines to use different weapons of the body and different close combat techniques to attack specific target areas.

Angles of Approach

Marines move anywhere within a 360-degree circle around the opponent to gain a tactical advantage. This circle provides access to different target areas of the opponent's body.

When facing an opponent, Marines move in a 45-degree angle to either side of the opponent. Moving at a 45-degree angle avoids an opponent's strike and puts Marines in the best position to attack the opponent. Marines should avoid being directly in front of an opponent during a confrontation. If a Marine is directly in front of an opponent, the opponent can rely on his forward momentum and linear power to seize the tactical advantage.

Movement

Marines must know how to move in all directions while maintaining the basic warrior stance. During any movement, the legs or feet should not be crossed. Once a movement is completed, the basic warrior stance should be resumed. Maintaining the basic warrior stance protects Marines and puts them in the proper position to launch an attack against an opponent.

Note: Before body movement begins, Marines turn their heads quickly to the new direction. The faster the head turns, the faster the body moves, and the quicker Marines attain visual contact with the opponent.

Forward to the Left. To move forward to the left from the basic warrior stance, Marines—

- Move the left foot forward at a 45-degree angle from the body (approximately 12 to 15 inches), keeping the toe pointed toward the opponent.
- Bring the right foot behind the left foot as soon as the left foot is in place. This returns Marines to the basic warrior stance.

Forward to the Right. To move forward to the right from the basic warrior stance, Marines—

- Move the right foot forward at a 45-degree angle from the body (approximately 12 to 15 inches).
- Bring the left foot, toe pointing toward the opponent, in front of the right foot as soon as the right foot is in place. This returns Marines to the basic warrior stance.

Backward to the Left. To move backward to the left from the basic warrior stance, Marines execute the forward movement in reverse. Marines—

- Move the left foot backward at a 45-degree angle from the body (approximately 12 to 15 inches), keeping the toe pointed toward the opponent.
- Bring the right foot behind the left foot as soon as the left foot is in place. This returns Marines to the basic warrior stance.

Backward to the Right. To move backward to the right from the basic warrior stance, Marines execute the forward movement in reverse. Marines—

- Move the right foot backward at a 45-degree angle from the body (approximately 12 to 15 inches).
- Bring the left foot, toe pointing toward the opponent, in front of the right foot as soon as the right foot is in place. This returns Marines to the basic warrior stance.

7. Balance and Off-Balancing

Balance

In any close combat situation, Marines must strive to maintain balance. The last place to be in a close combat situation is on the ground. Marines must maintain a strong base and a low center of balance, their feet must be a shoulder-width apart, and they must stay on their toes to enable quick movement.

Off-balancing

Marines use off-balancing techniques to control an opponent. These techniques are used to throw an opponent to the ground while Marines remain standing, or they are used to put Marines in a position for an offensive attack.

Off-balancing techniques use the opponent's momentum to move or throw him. For example, if the opponent is charging a Marine, the Marine pulls the opponent to drive him to the ground. Likewise, if the opponent is pulling a Marine, the Marine pushes the opponent to drive him to the ground.

Off-balancing techniques also rely on the power generated by the opponent. For example, during combat a Marine may be tired or outnumbered. Depending on the generated energy and momentum of the opponent, the Marine employs off-balancing techniques with very little effort and still provides effective results.

Because off-balancing techniques rely on the momentum and power generated by the opponent, these techniques are particularly effective for Marines who may be outsized by their opponent or lack their opponent's strength.

Angles of Off-balancing

There are eight angles or directions in which an opponent can be off-balanced: forward, rear, right, left, forward right, forward left, rear right, and rear left.

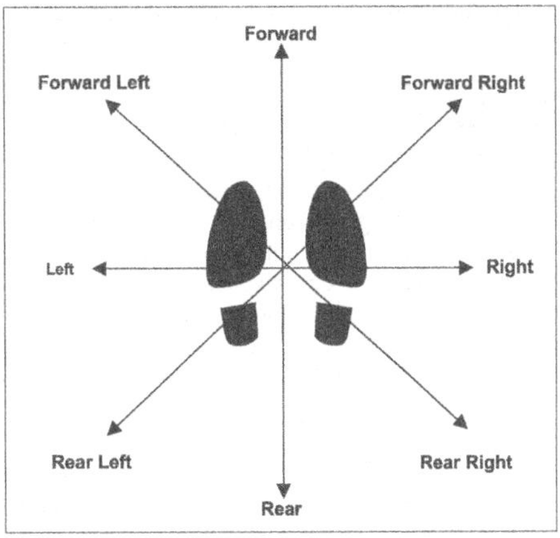

Note: The angles correspond to the Marine's perspective, not the opponent's. Forward, rear, right, and left are straight angles. Forward right, forward left, rear right, and rear left are considered quadrants that are at a 45-degree angle in either direction to the front or rear.

Off-balancing Techniques

Marines off-balance an opponent by pushing, pulling, or bumping the opponent with their hands, arms, or bodies.

To pull, Marines grab an opponent with their hands and drive him forcefully to one of the rear quadrants or to the right or left.

To push, Marines grab the opponent with their hands and drive him forcefully into one of the front quadrants or to the right or left.

Marines execute bumping in the same manner as pushing, but use their shoulders, hips, and legs instead of their hands to off-balance the opponent.

8. Falls

Marines may lose their balance or be thrown to the ground during encounters with an opponent. Marines use falling techniques to absorb the impact of a fall and to quickly return to their feet following an opponent's attack.

Whether falling or being thrown by an opponent, Marines strive to reduce the force of the impact, to prevent serious personal injury, and to increase the chances of survival. Falling techniques use the body's large muscles (back, thighs, buttocks) to protect vital organs and bones from injury and immobilization.

Front Fall

Marines execute a front fall to break a fall on the front. To execute the front fall, Marines—

Bend the elbows and place the palms facing out in a position to spread and absorb the impact of the fall.

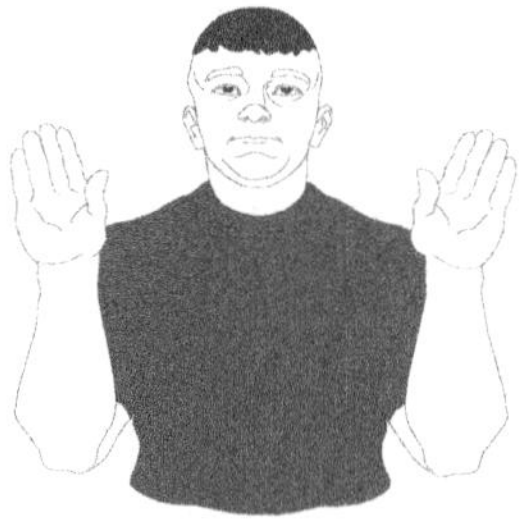

Fall forward, breaking the fall with the forearms and palms. The forearms and hands, down to the fingertips, should strike the ground simultaneously.

Offer resistance with the forearms and hands to keep the head raised off the ground.

Side Fall

Marines execute a side fall to break a fall on the side. To execute the side fall, Marines—

Bring the right arm across the body so the hand is next to the left shoulder with the palm facing inboard.

Fall to the side, breaking the fall with the right arm by slapping the ground and making contact from the shoulder or forearm down to the hand. At the same time, tuck the chin and keep the head raised off the ground. The chin should be tucked to the chest at all times to prevent whiplash.

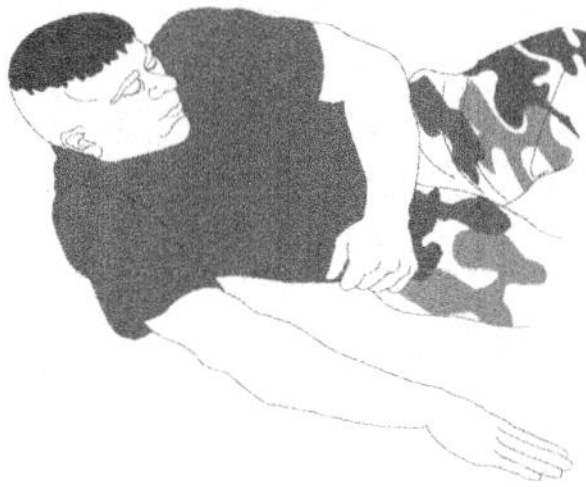

Stretch the right leg out to make contact with the ground and to distribute and absorb the impact. Bend the left leg, allowing the foot to make contact with the ground.

Back Fall

Marines execute a back fall to break the fall when being thrown or falling backward. To execute the back fall, Marines—

Cross the hands in front of the chest and tuck the chin.

Fall backward and slap the ground with the forearms and hands to absorb the impact of the fall and keep the head off the ground.

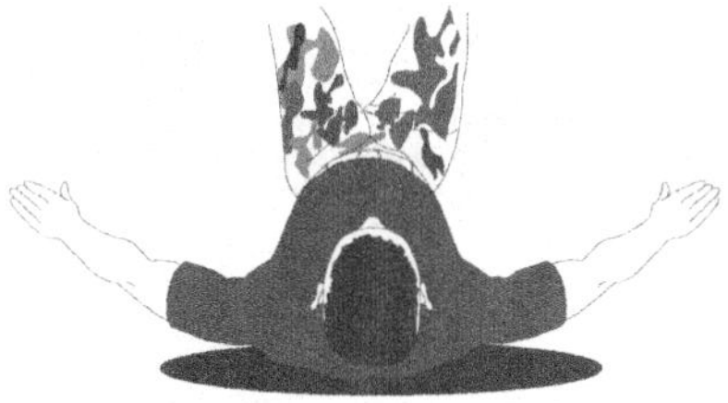

Forward Shoulder Roll

Marines use the forward shoulder roll to break a fall from an opponent's attack and to use the momentum of the fall to get back on their feet quickly. Ideally, Marines execute the forward shoulder roll to a standing position so they can continue fighting. To execute the forward shoulder roll, Marines—

Contact the ground with the back of the right forearm and upper arm. Tuck the chin into the chest.

Roll onto the right shoulder, rolling diagonally across the back to land on the left hip.

Slap the ground with the left arm, absorbing the impact from the shoulder to the hand, palm down.

Keep the left leg straight to absorb as much of the impact as possible. The right leg is bent and the foot hits flat on the ground.

Bend the left leg upon impact to push off with the left knee and leg to a squatting and then a standing position. Forward momentum should carry the Marine to a standing position.

CHAPTER 2

LETHAL AND NONLETHAL WEAPONS TECHNIQUES

This chapter describes all techniques for a right-handed person. However, all techniques can be executed from either side.

In drawings, the Marine is depicted in woodland camouflage utilities; the opponent is depicted without camouflage. In photographs, the Marine is depicted in woodland camouflage utilities; the opponent is depicted in desert camouflage utilities.

1. Bayonet Techniques

WARNING

During training, Marines must have bayonets sheathed. Marines use bayonet dummies to practice bayonet techniques. When practicing offensive and defensive bayonet techniques student-on-student, Marines use pugil sticks.

All Marines armed with a rifle carry a bayonet. The bayonet is an effective weapon if Marines are properly trained in offensive and defensive bayonet techniques. An offensive attack, such as a thrust, is a devastating attack that can quickly end a fight. Defensive techniques, such as the block and parry, can deter the opponent's attack and allow Marines to regain the initiative. Through proper training, Marines develop the courage and confidence required to effectively use a bayonet to protect themselves and destroy the enemy. In situations where friendly and enemy troops are closely mingled and rifle fire and grenades are impractical, the bayonet becomes the weapon of choice.

Holding the Rifle

To execute bayonet techniques, Marines hold the rifle in a modified basic warrior stance. All movement begins and ends with the basic warrior stance. To hold the rifle, Marines—

Use an overhanded grasp to grab the small of the rifle's stock. Use an underhanded grasp to grab the handguards of the rifle.

Lock the buttstock of the rifle against the hip with the right forearm.

Orient the blade end of the rifle toward the opponent.

Offensive Bayonet Techniques

Straight Thrust. Marines use the straight thrust to disable or kill an opponent. It is the most deadly offensive technique because it causes the most trauma to an opponent. Target areas are the opponent's throat, groin, or face. The opponent's chest and stomach are also excellent target areas if not protected by body armor or combat equipment. To execute the straight thrust, Marines—

Lift the left leg and lunge forward off the ball of the right foot while thrusting the blade end of the weapon forward, directly toward the opponent.

Retract the weapon and return to the basic warrior stance.

Slash. Marines use the slash to kill an opponent or to create an opening in his defense. The target area is the opponent's neck. To execute the slash, Marines—

Extend the left hand back toward the left shoulder.

Thrust the left hand forward and swing it to the right, bring the right hand back toward the hip, and turn the cutting edge of the blade toward the opponent's neck. The movement is a slashing motion so the blade cuts across the opponent's neck.

Horizontal Buttstroke. Marines use the horizontal buttstroke to weaken an opponent's defenses, to cause serious injury, or to set him up for a killing blow. Target areas are the opponent's head, neck, and legs. To execute the horizontal buttstroke, Marines—

Step forward with the right foot and drive the right hand forward. Rotate the hips and shoulders into the strike. Move the left hand back toward the left shoulder.

Strike the opponent with the butt of the weapon.

Vertical Buttstroke. Marines use the vertical buttstroke to weaken an opponent's defenses, to cause serious injury, or to set him up for a killing blow. Target areas are the opponent's groin and face. To execute the vertical buttstroke, Marines—

Step forward with the right foot and drive the right hand straight up.

Pull the left hand back over the left shoulder.

Strike the opponent with the butt of the weapon.

Smash. Marines use the smash as a follow-up technique to the horizontal or vertical buttstroke, particularly if they missed the target. The target area is the opponent's head. To execute the smash following a buttstroke, Marines—

Step forward with the right foot and place the blade end of the weapon over the left shoulder and elevate the right elbow above the shoulders.

Drive the arms straight forward, striking the opponent with the butt of the weapon.

Defensive Bayonet Techniques

Parry. Marines use a parry as a defensive technique to redirect or deflect an attack. A parry is a slight redirection of a linear attack by an opponent; e.g., a straight thrust or a smash. To execute the parry, Marines—

Use the bayonet end of the rifle to redirect the barrel or bayonet of the opponent's weapon.

Lock the weapon against the hip with the right forearm.

Rotate to the right or left, moving the bayonet end of the rifle to parry the opponent's attack. Rotation is generated from the hips.

Redirect or guide the opponent's weapon away from the body by exerting pressure against the opponent's weapon.

High Block. Marines execute a high block against a vertical attack coming from high to low. To execute the high block, Marines—

Thrust the arms up forcefully at approximately a 45-degree angle from the body. The weapon should be over the top of the head and parallel to the ground. The elbows are bent, but there is enough muscular tension in the arms to absorb the impact and deter the attack.

Low Block. Marines execute the low block

against a vertical attack coming from low to high. To execute the low block, Marines—

Thrust the arms down forcefully at approximately a 45-degree angle from the body. The weapon

should be at or below the waist and parallel to the ground. The elbows are bent, but there is enough muscular tension in the arms to absorb the impact and deter the attack.

Left and Right Block. Marines execute a left or right block against a horizontal buttstroke or a slash. To execute the left or right block, Marines—

Thrust the arms forcefully to the right or left, holding the rifle vertically in the direction of the attack. The elbows are bent, but there is enough muscular tension in the arms to absorb the impact and deter the attack.

Counter Action Following the Block. After deflecting an opponent's attack with a block, Marines counter with a slash or a horizontal buttstroke to regain the initiative. However, the objective in any bayonet fight is to thrust forward with the blade end of the weapon to immediately end the fight.

Group Strategy

On occasion, Marines may engage an opponent as a member of a group or numerous opponents by one's self or as a member of a group. By combining bayonet fighting movements and simple strategies, Marines can effectively overcome their opponent or opponents.

Offensive Strategy: Two Against One. If two bayonet fighters engage one opponent, the fighters advance together.

Fighter 1 engages the opponent while fighter 2 swiftly and aggressively attacks the opponent's exposed flank and destroys the opponent.

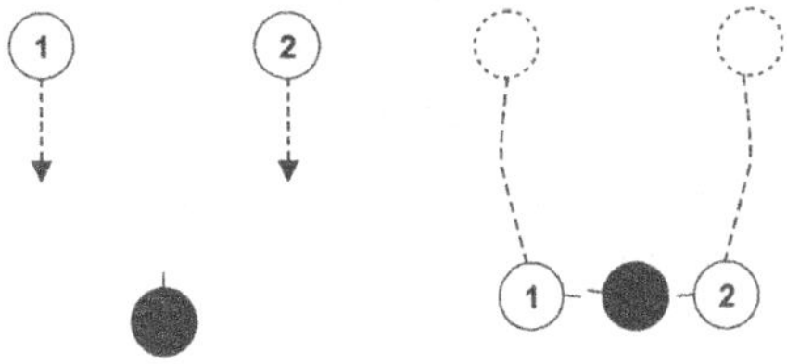

Offensive Strategy: Three Against Two. If three bayonet fighters engage two opponents, the fighters advance together keeping their opponents to the inside.

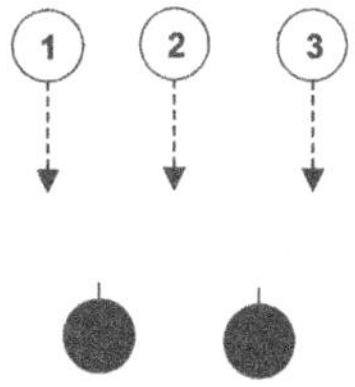

Fighters 1 and 3 engage opponents. Fighter 2 attacks the opponent's exposed flank engaged by fighter 1 and destroys the opponent.

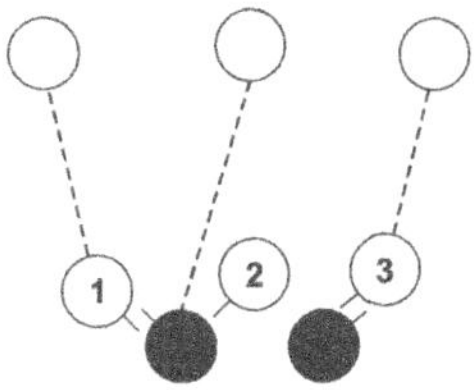

Fighters 1 and 2 turn and attack the exposed flank of the opponent engaged by fighter 3 and destroy the opponent.

Defensive Strategy: One Against Two. If a fighter is attacked by two opponents, the fighter immediately positions himself at the flank of the nearest opponent and keeps that opponent between himself and the other opponent.

Using the first opponent's body as a shield against the second opponent, the fighter destroys the first opponent quickly before the second opponent moves to assist.

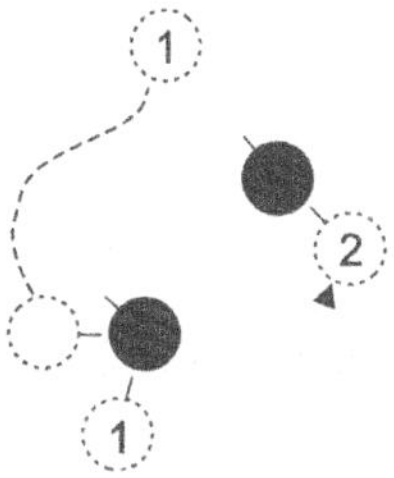

Then, the fighter engages and destroys the second opponent.

Defensive Strategy: Two Against Three. If two fighters are attacked by three opponents, the fighters immediately move to the opponent's flanks.

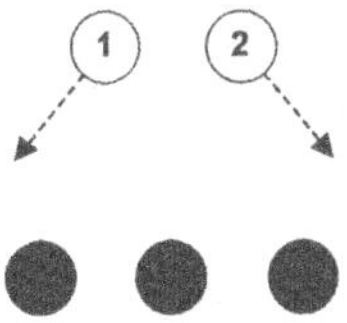

Fighters 1 and 2 quickly attack and destroy their opponents before the third opponent closes in.

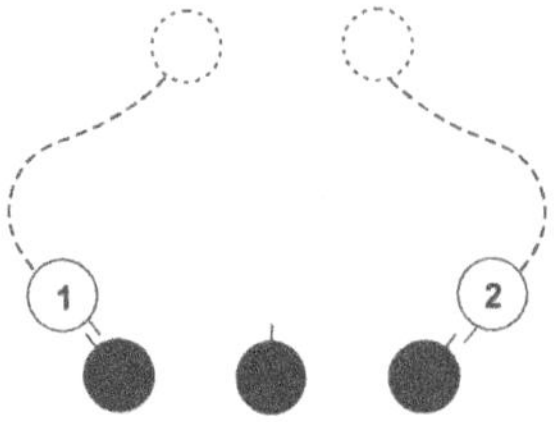

Fighter 1 engages the third opponent while fighter 2 attacks the opponent's exposed flank and destroys the opponent.

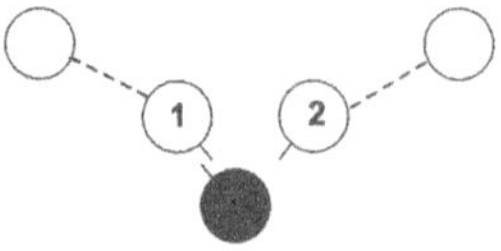

2. Nonlethal Rifle and Shotgun Retention Techniques

Most Marines are armed with the M16A2 service rifle. Marines are taught to keep their weapon with them at all times. Marines must be constantly alert to their surroundings and the people moving in and around their environment. Marines may be confronted by an individual who tries to take their weapons. If this happens, Marines should not struggle with the individual. To retain positive control of their weapons, Marines must understand and apply weapons retention techniques, otherwise known as armed manipulation. The following techniques can be used with either the rifle or the shotgun.

Blocking Technique

To execute a blocking technique, Marines—

Stand in a defensive posture.

Use the weapon to block the opponent by thrusting it out firmly, with the elbows bent. Do not try to hit the opponent with the rifle. The rifle is used as a barrier.

Technique if Opponent Grabs Weapon Underhanded

Usually an opponent will try to grab the weapon or block it as an instinctive action. If the opponent uses an underhand grab to seize the handguards of the weapon, Marines—

Trap the opponent's closest finger(s) above the knuckle with the thumb so he cannot release his grip.

Apply bone pressure on the opponent's finger to initiate pain compliance.

Rotate the barrel of the weapon up or down quickly while maintaining pressure on the opponent's hand. At the same time, quickly pivot to off-balance the opponent.

Technique if Opponent Grabs Weapon Overhanded

If the opponent uses an overhand grab to seize the weapon overhanded, Marines—

Trap the opponent's finger to hold his hand in place.

Rotate the barrel to place it across the opponent's forearm and apply downward pressure. This action is similar to an armbar.

Technique if the Opponent Grabs the Muzzle

If the opponent grabs the muzzle of the weapon, Marines—

Rotate the muzzle quickly in a circle motion. Slash downward with the muzzle to release the opponent's grip.

Butt Strikes

Strikes with the butt of the weapon control or ward off an attacker. During any of the retention techniques, Marines use the heel or cutting edge of the weapon to deliver butt strikes to the inside or the outside of the opponent's thigh. The inside butt strike targets the femoral nerve as illustrated in the photo to the left.

The outside butt strike targets the peroneal nerve as illustrated in the photo to the left. Strikes can be made to the outside or inside of the thighs. If a strike to one side of the thigh misses, Marines follow back through with the butt of the weapon on the other side of the thigh.

Off-balancing Techniques

Marines apply off-balancing techniques to throw an opponent to the ground and retain possession of the weapon.

If the opponent grabs the weapon and pushes, Marines should not push on the weapon. They should—

Move with the momentum and movement of the

opponent by pivoting in the direction of the movement by stepping back.

Throw the opponent to the ground with a quick jerking movement by lowering the muzzle and swinging the butt of the weapon.

If the opponent grabs the weapon and pulls, Marines—

Step on the opponent's foot and push forward to off-balance him and drive him to the ground.

Sweep the opponent's feet out from under him by hooking his leg with the leg and kicking backward.

3. Nonlethal Handgun Retention Techniques

Many Marines are armed with the M9 service pistol. Marines must keep their weapons in their possession at all times. Marines must be constantly alert to their surroundings and the people moving in and around the environment. Marines may be confronted by an individual who tries to take their weapons. To retain positive control of the weapon, Marines must understand and apply handgun retention techniques.

Blocking Technique

Marines perform the following blocking techniques if an opponent attempts to grab their pistol in the holster. Marines—

Place the body between the weapon and the opponent by immediately pivoting so the weapon is away from the opponent.

Step back and away from the opponent while placing the hand on the pistol grip.

Extend the left hand and block, deflect, or strike the opponent's arm.

Armbar Technique

Marines use the armbar technique when an opponent uses his right hand to grab their pistol in the holster. To execute the armbar technique, Marines—

Trap the opponent's right hand by grasping the opponent's wrist or hand with the right hand and applying pressure against the body.

Step back with the right foot and pivot sharply to the right to be next to the opponent. Always pivot in a direction that keeps the weapon away from the opponent.

Straighten the opponent's arm to apply an armbar. The arm should be straight across the torso.

Continue pivoting to the right while pulling back on the opponent's shoulder. This action may break the opponent's arm.

Place the left hand across the opponent's face and apply pressure back and down to take the oppo-

nent to the ground. Pressure applied beneath the nose or on the trachea is equally effective.

Wristlock Technique

Marines use the wristlock technique when an opponent grabs their pistol while it is in the holster with his right hand. To execute the wristlock technique, Marines—

Grasp the opponent's wrist or hand with the right hand and apply pressure against the body.

Step back and away from the opponent and pivot to the right so the weapon is away from the opponent. Always pivot in a direction that keeps the weapon away from the opponent.

Reach over the opponent's arm with the left hand and grab his hand, applying pressure against his arm with the left forearm. Execute a wristlock.

Incorporate the second hand into the wristlock and, stepping back with the left foot, pivot to the left.

Execute a two-handed wristlock by exerting downward pressure with the thumb and rotating his hand to the left.

Continue to pivot to off-balance the opponent and drive him to the ground.

Note: If the opponent grabs the pistol with his left hand, Marines execute the wristlock with one hand and step in toward the opponent, rather than

away from the opponent.

Softening Techniques

Handgun retention techniques use softening techniques applied to pressure points. Bone pressure

and strikes with the hands (i.e., hammer fist), knees, and feet are also effective softening techniques.

Pressure Points. Marines use pressure point techniques to get the opponent to loosen his grip. Marines use their finger tips to apply pressure to the webbing between the index finger and thumb, the jugular notch, and the brachial plexus tie in. The following figure illustrates pressure applied to the brachial plexus tie in.

Bone Pressure. Bone pressure is the application of pressure on a bone against a hard object to initiate pain compliance. To apply bone pressure, Marines use their hand to trap the opponent's hand on the weapon. Marines apply a slow, steady pressure to the opponent's hand and fingers until his grip is softened or he releases his hold.

Strikes. If it is difficult to apply a retention technique, Marines employ strikes or kicks to force the opponent to loosen his grip. Strikes to the

eyes, the arms (radial nerve), or shoulder (brachial plexus tie in) soften the opponent's grip on the weapon.

Kicks and knee strikes to the peroneal nerve, the femoral nerve, or the groin are effective because the opponent is typically unprepared to counter the strike.

Stomping on the top of the opponent's foot may distract him or loosen his grip on the weapon.

4. Firearm Disarmament Techniques

Marines use firearm disarmament techniques during a close-range confrontation if they are un-

armed and the opponent has a firearm (pistol). These techniques are equally effective if Marines are armed but do not have time to withdraw and present the weapon. The goal of firearm disarmament techniques is to gain control of the situation

so Marines gain the tactical advantage. The goal is not necessarily to get control of the opponent's weapon.

Pistol to the Front

This technique is used when Marines are unarmed and the opponent has a pistol pointing at their front (e.g., head, chest). The technique is the same if the opponent sticks the pistol under the Marine's chin. To execute the counter when an opponent is pointing a pistol toward the front of a Marine, Marines—

Place the hands close to the weapon, about chest high, palms out.

Use the left hand to grab the opponent's forearm and push the opponent's hand with the pistol away to clear the body from in front of the weapon. At the same time, rotate the right shoulder back to clear the body from the weapon.

Maintain control of the opponent's arm.

Grasp the weapon with the right hand by placing the thumb underneath the pistol and the fingers over top of the pistol.

Keep the right hand wrapped tightly around the muzzle and quickly rotate the pistol in the opponent's hand so the muzzle is facing the opponent.

Grasp and pull the opponent's wrist or forearm away from the body while rotating the weapon.

Rotate the weapon toward the opponent while pulling it up and back and out of the opponent's grasp.

Pistol to the Rear

This technique is performed when Marines are unarmed and the opponent has a pistol pointing to the back of the Marine's head. To execute the counter when the opponent is pointing a pistol to the rear, Marines—

Place the hands close together about chest high, palms out.

Step back with the left foot, pivoting on the right foot so the side is against the opponent's front. This action clears the body from the weapon's line of fire. Keep the left hand up.

Pivot on the left foot to face the opponent and, at the same time, raise the left elbow and reach over the top of the opponent's arm with the left arm.

Wrap the left arm tightly around the opponent's arm above his elbow to control it.

Push on the opponent's shoulder with the right hand while pulling up with the left arm to achieve an armbar. This action releases the opponent's grip on the weapon. If necessary, the opponent can be taken to the ground with a leg sweep.

Note: To execute this technique, the weapon must be close to or touching the rear of the Marine. If the weapon is too far away from the body, this technique would be difficult to execute or it would be ineffective.

CHAPTER 3

HAND-HELD WEAPONS

This chapter describes all techniques for a right-handed person. However, all techniques can be executed from either side.

In drawings, the Marine is depicted in woodland camouflage utilities; the opponent is depicted without camouflage. In photographs, the Marine is depicted in woodland camouflage utilities; the opponent is depicted in desert camouflage utilities.

Marines must know how to defend against attacks when an opponent is either unarmed or armed with a held-held weapon. This chapter addresses the combative use of knives, specific weapons of opportunity, and sticks. However, virtually anything can be used as a hand-held weapon.

1. Fundamentals of Knife Fighting

Marines must be trained in knife fighting techniques. Marines experienced in offensive knife techniques can cause enough damage and massive trauma to stop an opponent. When engaged against each other, experienced knife fighters employ various maneuvers and techniques that are specific to knife fighting. Seldom, if ever, will Marines engage an opponent in a classical knife fight.

Note: When armed with a rifle, Marines are issued a bayonet. When armed with a pistol, Marine are issued a combat knife.

Angles of Attack

There are six angles from which an attack with a knife can be launched:

- Vertical strike coming straight down on an opponent.
- Forward diagonal strike coming in at a 45-degree angle to the opponent.
- Reverse diagonal strike coming in at a 45-degree angle to the opponent.
- Forward horizontal strike coming in parallel to the ground.
- Reverse horizontal strike coming in parallel to the ground.
- Forward thrust coming in a straight line to the opponent.

Target Areas of the Body

During any confrontation, the parts of the opponent's body that are exposed or readily accessible will vary. The goal in a knife fight is to attack the body's soft, vital target areas that are readily accessible (e.g, the face, the sides and front of the neck, the lower abdomen [or groin]).

Neck. Carotid arteries, located on either side of the neck, are good target areas because they are not covered by body armor or natural protection.

Lower Abdomen (or Groin). The lower abdomen (or groin region) is a good target area because it is not covered by body armor.

Heart. The heart, if not covered by body armor, is an excellent target which, if struck, can prove fatal in a matter of seconds or minutes.

Secondary Targets. There are secondary target areas that will cause substantial bleeding if an artery is severed. These target areas are not immediately fatal, but can become fatal if left unattended. Attacks to—

- The legs can cause a great deal of trauma and prove fatal. For example, the femoral artery located on the inside of the thigh is a large artery which, if cut, will cause extensive blood loss.
- The brachial artery, located between the biceps and triceps on the inside of the arm, can cause extensive bleeding and damage.
- The arm's radial and ulnar nerves can cause extensive bleeding and damage.

Movement

Marines can move anywhere within a 360-degree circle around the opponent. This allows accessibility to different target areas of the opponent's body. Marines should avoid being directly in front of an opponent because the opponent can rely on his forward momentum to seize the tactical advantage. If Marines face an opponent, movement is made in a 45-degree angle to either side of the opponent. This angle avoids an opponent's strike and places Marines in the best position to attack an opponent.

Wearing the Combat Knife

Marines must wear the combat knife where it is easily accessible and where it can best be retained. It is recommended the combat knife be worn on the weak side hip, blade down, sharp edge facing forward. Marines can place it behind the magazine pouch where it is easily accessible to them, but not easily grabbed by an opponent.

Grip

The grip on the knife should be natural. Marines grasp the knife's grip with the fingers wrapped around the grip naturally as it is pulled out of its sheath. This is commonly known as a hammer grip. The blade end of the knife is always facing the opponent.

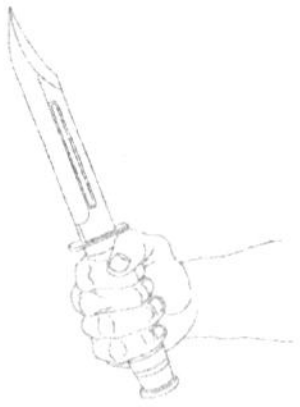

Stance

Marines use the basic warrior stance as the foundation for knife techniques. The left hand is a vertical shield that protects either the ribs or the head and neck. The right elbow is bent with the blade pointing forward toward the opponent's head. This position serves as an index point, where all techniques are initiated.

Principles of Knife Fighting

The following are key principles of knife fighting:

- Execute movements with the knife blade within a box, shoulder-width across from the neck down to the waistline. The opponent has a greater chance of blocking an attack if the blade is brought in a wide, sweeping movement to the opponent.
- Close with the opponent, coming straight to the target.
- Move with the knife in straight lines.
- Point the knife's blade tip forward and toward the opponent.
- Apply full body weight and power in each of the knife techniques. Full body weight should be put into the attack in the direction of the blade's movement (slash or thrust).
- Apply constant forward pressure with the body and blade to keep the opponent off-balanced.

2. Knife Fighting Techniques

Slashing Techniques

Marines use slashing techniques to close with an enemy. Slashing techniques distract the opponent or damage the opponent so Marines can close in. Typically, Marines target the opponent's limbs, but any portion of the body that is presented can become a target.

Vertical Slash Technique. The vertical slash follows a vertical line straight down through the target. To execute the vertical slash, Marines—

Thrust the right hand out and bring the weapon straight down on the opponent, continuing to drag the knife down through the opponent's body.

Maintain contact on the opponent's body with the blade of the knife.

Forward Slash Technique. The forward slash follows a straight line in a forehand stroke, across the target areas of either the neck (high diagonal slash) or abdominal region (low horizontal slash). To execute the forward slash, Marines—

Extend the right hand while simultaneously rotating the palm up until the knife blade makes contact with the opponent.

Snap or rotate the wrist through the slashing motion to maximize blade's contact with the opponent.

Drag the knife across the opponent's body, from right to left, in a forehand stroke. The movement ends with the forearm against the body and the knife at the left hip with its blade oriented toward the opponent.

Reverse Slash Technique. The reverse slash is a follow-up technique to a forward attack. It allows Marines both a secondary attack and the ability to resume the basic warrior stance. The reverse slash follows a straight line in a backhand stroke, across the target areas of either the neck (high diagonal slash) or abdominal region (low horizontal slash). To execute the reverse slash, Marines—

Extend the right hand while simultaneously rotating the palm down until the knife blade makes contact with the opponent.

Snap or rotate the wrist through the slashing motion to maximize the blade's contact with the opponent.

Drag the knife across the opponent's body, from left to right, in a backhand stroke. Maintain contact on the opponent's body with the blade of the knife.

Thrusting Techniques

The primary objective of knife fighting is to insert the blade into an opponent to cause extensive damage and trauma. This is done with a thrusting technique. Thrusting techniques are more effective than slashing techniques because of the damage they can inflict. However, Marines use slashing techniques to close with the enemy so that they are closer to the opponent, which allows them to use the thrusting technique.

Vertical Thrust. The thrusting motion follows a vertical line straight up through the target (low into the abdomen region or high into the neck). To execute the vertical thrust, Marines—

Thrust the right hand toward the target, inserting the knife blade straight into the opponent.

Pull the knife out of the opponent.

Forward Thrust. The forward thrust follows a straight line straight into the opponent's neck (high thrust) or abdominal region (low thrust). To execute the forward thrust, Marines—

Thrust the right hand, palm down, toward the target, inserting the knife blade straight into the opponent.

Rotate the palm up once the knife is inserted to twist the blade.

Drop the right elbow and bring the knife to the opposite side of the opponent's body from where it was inserted. At the same time, rotate the hips and shoulders downward to bring body weight to bear on the attack.

Reverse Thrust. The reverse thrust is a follow-up technique to a forward attack. It allows Marines both a secondary attack and the ability to resume the basic warrior stance. The reverse thrust follows a horizontal line straight into the opponent's neck (high thrust) or abdominal region (low thrust). To execute the reverse thrust, Marines—

Bend the right arm, crossing the arm to the left side of the body.

Thrust the right hand, palm up, toward the target, inserting the knife blade straight into the opponent.

Rotate the palm down to twist the blade once the knife is inserted.

Bring the knife to the opposite side of the opponent's body from where it was inserted.

3. Weapons of Opportunity

During an unarmed close combat situation, Marines use their bodies as weapons, but they should be ready and able to use anything around them as a weapon. For example, Marines could throw sand or liquid in an opponent's eyes to temporarily impair his vision or smash the opponent's head with a rock or helmet. Marines must use whatever means are available and do whatever it takes to take control of the situation and to win, or they face the possibility of losing their lives. Some weapons of opportunity are discussed in the following subparagraphs.

Entrenching Tool

An entrenching tool (E-tool) is commonly carried by Marines. It can be an excellent weapon, especially when sharpened. Marines can use the E-tool to block, slash, and thrust at an opponent.

Tent Pole and Pins

Marines can use tent poles and pins to block, strike, or thrust at an opponent.

Web Belt

Marines can stretch a web belt between their hands to block attacks by an opponent.

Battlefield Debris

Marines can use debris on the battlefield (e.g., sticks, glass, a sharp piece of metal) to cut, slash, or stab an opponent. They can also use other types of debris such as shovel or ax handles, boards, metal pipes, or broken rifles to strike an opponent or apply a choke.

Helmet

A helmet can be used to strike an opponent on an unprotected area like the head and face. Grasp the rim of the helmet and thrust the arms forward, striking the opponent with the top of the helmet.

4. Fundamentals of Combative Stick

On the battlefield, Marines must be ready and able to use anything as a weapon. They must learn and be able to use techniques that can be employed with most weapons of opportunity. Among these techniques are combative stick techniques. Combative stick techniques can be used with a stick, a club, a broken rifle, an E-tool, or even a web belt.

Angles of Attack

There are six angles from which an attack with a hand-held weapon can be launched:

- Vertical strike coming straight down on an opponent.
- Forward diagonal strike coming in at a 45-degree angle to the opponent.
- Reverse diagonal strike coming in at a 45-degree angle to the opponent.
- Forward horizontal strike coming in parallel to the ground.
- Reverse horizontal strike coming in parallel to the ground.
- Forward thrust coming in a straight line to the opponent.

Grip

Grasp the stick about 2 inches from its base.

Stance

The basic warrior stance serves as the foundation for combative stick techniques. The left hand becomes a vertical shield that protects the ribs or the head and neck. Depending on how heavy the weapon is, it should be held at a level approximately shoulder height.

Movement

Movement during combative stick techniques is the same as it is for other close combat techniques. Marines can move anywhere within a 360-degree circle around the opponent. This allows accessibility to different target areas of the opponent's body and gains a tactical advantage.

Marines should avoid being directly in front of an opponent because the opponent can rely on his forward momentum to seize the tactical advantage. If Marines face an opponent, movement is made in a 45-degree angle to either side of the opponent. This angle avoids an opponent's strike and places Marines in the best position to attack an opponent.

5. Combative Stick Techniques

Strikes

Strikes are intended to inflict as much damage on an opponent as possible. Striking techniques apply to a weapon of opportunity such as a stick, a tent pole, a club, a broken rifle, an E-tool, or a pipe.

Vertical Strike. To execute the vertical strike, Marines—

Bend the right arm, extending the weapon over the back of the right shoulder.

Rotate the forearm straight down off the elbow to bring the weapon down on the opponent.

Rotate the hips and shoulders forcefully toward the opponent.

Follow through with the strike by allowing the weight of the weapon to go through the target area of the body.

Forward Strike. To execute a forward strike, Marines—

Step forward with the left foot in the direction of the strike.

Bend the right arm with the elbow extending out to the right and the weapon extended over the right shoulder.

Rotate the forearm to the right of the elbow to bring the weapon down onto the opponent. At the same time, forcefully rotate the hips and shoulders toward the opponent.

Follow through with the strike by allowing the weight of the weapon to go through the target area of the body.

Reverse Strike. The reverse strike is a follow-up technique to a forward strike. It allows Marines both a secondary attack and the ability to resume the basic warrior stance.To execute a reverse strike, Marines—

Step forward with the right foot in the direction of the strike.

Bend the right arm with the hand near the left shoulder. The weapon is extended over the left shoulder.

Rotate the forearm to the right of the elbow to bring the weapon down onto the opponent. At the same time, forcefully rotate the hips and shoulders toward the opponent.

Follow through with the strike by allowing the weight of the weapon to go through the target area of the body.

Forward Thrust. To execute the forward thrust, Marines—

Grasp the stick with the left hand, palm up, in a position where the stick can be controlled with two hands.

Lift the left leg and lunge forward off the ball of the right foot. At the same time, thrust the end of the weapon directly toward the opponent by thrusting both hands forward in a straight line.

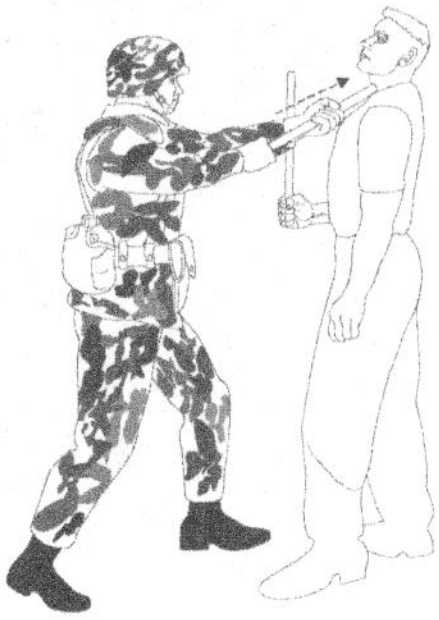

6. Blocking Techniques

A block is meant to deter or deflect an attack by an opponent. A block sets up Marines for a follow-on attack against the opponent. Blocks are executed by deflecting, rather than hitting or following through like a strike.

Blocks Against Unarmed Attacks

To block against an unarmed attack, Marines—

Step forward at a 45-degree angle with either the right or left foot. This moves the body out of the line of attack.

Raise the left arm and block the strike with the meaty portion of the forearm.

Employ the stick by one of two techniques:

- Strike the opponent with the stick.
- Use the stick to block on two points of contact. When the stick is used to block, it serves as an incidental strike. This technique is only used if Marines have closed with the opponent and are inside his strike.

Blocks Against Armed Attacks

Block for a Vertical Strike. To execute the block for a vertical strike, Marines—

Step forward with the left foot at a 45-degree angle to the left. This moves the body out of the line of attack.

Block on two points of contact to disperse the impact of the attack:

- Block the opponent's stick by positioning the stick so it is perpendicular to the opponent's stick. If the stick is not perpendicular to the opponent's stick, the stick can slide through and make contact on the Marine.
- Block the opponent's wrist or forearm with the back of the left forearm.

Note: Use the stick to block the opponent's arm if closer to the opponent. It is the same movement as blocking with the arm, except the opponent's arm is blocked with both the stick and the arm.

Block for a Forward Strike. To execute the block for a forward strike, Marines—

Step forward with the left foot at a 45-degree angle to the left. This moves the body out of the line of attack and inside the opponent's strike.

Block on two points of contact to disperse the impact of the attack:

- Block the opponent's wrist or forearm with the meaty portion of the back of the left forearm.
- Strike the opponent's attacking biceps with the stick.

Block for a Reverse Strike. To execute the block for a reverse strike, Marines—

Step forward at a 45-degree angle with the right foot to the right. This moves the body out of the line of attack and inside the opponent's strike.

Block on two points of contact to disperse the impact of the attack:

- Block the opponent's stick by positioning the stick so it is perpendicular to the opponent's stick.

Block the opponent's forearm with the meaty portion of the left forearm.

Note: If closer to the opponent, block his triceps with the back of the left forearm and strike his forearm with the stick.

7. Unarmed Against Hand-Held Weapons

If Marines are engaged against an opponent with a knife, a stick, or some other weapon of opportunity, they must establish and maintain an offensive mindset, not a defensive mindset. Their survival depends on it. They cannot afford to think about getting cut or hurt.

Angles of Attack

Before Marines can learn to block or counter an attack with a hand-held weapon (e.g., knife, stick), they must know from what angle the opponent is attacking. There are six angles from which an opponent will typically attack with a hand-held weapon:

- Vertical strike coming in straight down on the Marine.
- Forward diagonal strike coming in at a 45-degree angle to the Marine.
- Reverse diagonal strike coming in at a 45-degree angle to the Marine.
- Forward horizontal strike coming in parallel to the ground.
- Reverse horizontal strike coming in parallel to the ground.
- Forward thrust coming in a straight line to the Marine.

Blocks

If the opponent has a hand-held weapon (e.g., knife or stick), Marines parry the opponent's hand or arm to block the attack.

Basic Block. To execute the basic block technique, Marines—

Step forward at a 45-degree angle to move out of the line of the attack. Always step in the direction of the strike.

Thrust the forearms forward, hands up, against the opponent's attacking arm. Contact is made on the opponent's arm with the backs of the forearms.

Block for a Vertical Strike. To execute the block against a vertical strike, Marines—

Step forward with the left foot at a 45-degree angle to the left to move out of the line of attack.

Thrust the forearms forward, hands up, against the outside of the opponent's attacking arm.

Block for a Forward Strike. To execute the block against a forward diagonal or forward horizontal strike, Marines—

Step forward with the left foot inside the opponent's attacking arm.

Block the attack with both arms bent so the forearms make contact with the opponent's biceps and forearm.

Block for a Reverse Strike. To execute the block against a reverse diagonal or reverse horizontal strike, Marines—

Step forward with the right foot to the outside of the opponent's attacking arm.

Block the attack with both arms bent so the forearms make contact with the opponent's triceps and forearm.

Block for a Forward Thrust. To execute the block against a forward thrust, Marines—

Bend at the waist, move the hips backward, and jump backward with both feet to move away from the attack. This action is known as "hollowing out."

Hollow out and block the attack with the arms bent and hands together on top of the attacking arm.

Overlap the hands slightly so one thumb is on top of the other hand's index finger. The other thumb should be under the other hand's index finger.

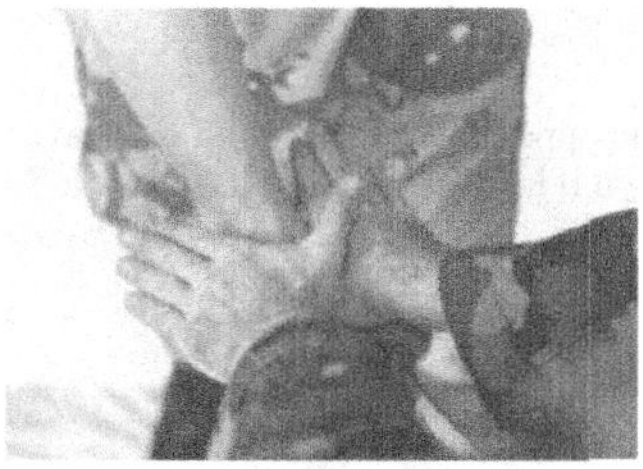

8. Counters to Hand-Held Weapon Attacks

Principles of Counters

A counter is used to control the situation to regain the tactical advantage and end the fight. Regardless of the type of weapon or angle of attack, the following principles apply to countering the attack with a hand-held weapon:

- Move out of the line of attack. Movement is executed in a 45-degree angle forward to the left or right.
- Block the attack.

Note: The first two actions are taken simultaneously.

- Take control of the weapon by controlling the hand or arm that is holding the weapon. Never attempt to grab the opponent's weapon.
- Execute the appropriate follow-up to end the fight; e.g., strikes, joint manipulations, throws, or takedowns (see chap. 8). Marines should continue to attack the opponent until the fight ends.

Counter Techniques

There are two techniques that can be used to counter any armed attack: forward armbar counter and reverse armbar counter. These techniques can be used to counter a vertical attack, a forward diagonal strike, or a forward horizontal strike. With minor variations, the same techniques are used to counter reverse strikes. A third technique, the bent armbar counter, is used to counter a vertical attack.

Forward Armbar Counter. To execute the forward armbar counter to an attack coming from a forward strike, Marines—

Step forward with the left foot inside the opponent's attacking arm.

Block the attack with both arms bent so the forearms make contact with the opponent's biceps and forearm.

Slide the left arm over the opponent's forearm and wrap the arm tightly around his arm, trapping the opponent's attacking arm between the biceps and torso.

Place the right hand on the opponent's shoulder or upper arm to further control his arm and to effect an armbar.

Execute an armbar and continue to exert steady pressure against the arm.

Reverse Armbar Counter. To execute the reverse armbar counter to an attack coming from a forward strike, Marines—

Step forward with the left foot inside the opponent's attacking arm.

Block the attack with both arms bent so the forearms make contact with the opponent's biceps and forearm.

Control the opponent's arm with the left hand and pivot to the right so the back is against the opponent's side. Immediately slide the right arm over the opponent's biceps and wrap the arm tightly around his arm, trapping the opponent's attacking arm between the biceps and torso.

Grasp the opponent's wrist with the left hand and twist his thumb away from the body.

Take the opponent to the ground with an armbar takedown if he does not drop the weapon from the wristlock.

Bent Armbar Counter. This counter is particularly effective against a vertical attack. To execute the bent armbar counter, Marines—

Step forward with the left foot inside the opponent's attacking arm.

Block the attack with both arms bent so the forearms make contact with the opponent's biceps and forearm.

Grasp the opponent's forearm with the left hand. At the same time, slide the right arm underneath the opponent's triceps and grasp the opponent's forearm or wrist with the right hand.

Apply downward pressure with the hands against the opponent's forearm to off-balance the opponent.

CHAPTER 4

STRIKES

This chapter describes all techniques for a right-handed person. However, all techniques can be executed from either side.

In drawings, the Marine is depicted in woodland camouflage utilities; the opponent is depicted without camouflage. In photographs, the Marine is depicted in woodland camouflage utilities; the opponent is depicted in desert camouflage utilities.

Strikes are unarmed individual hitting techniques. Strikes use the hands, elbows, knees, feet, and, in some instances, other parts of the body as personal weapons. Marines must know how to execute strikes effectively. They must also know how to counter strikes from an opponent.

1. Principles of Punches

Muscle Relaxation

Muscle relaxation is crucial when executing punches. The natural tendency in a fight is to tense up, which results in rapid fatigue and decreased power generation. Marines who remain relaxed during a close combat situation generate greater speed, which results in greater generation of power. Relaxing the forearms generates speed and improves reaction time. At the point of impact, Marines clench the fist to cause damage to the opponent and avoid injury to the wrist and hand.

Weight Transfer

Weight transfer is necessary to generate power in a punch. Marines accomplish this by—

- Rotating their hips and shoulders into the attack.
- Moving their body mass straight forward or backward in a straight line.
- Dropping their body weight into an opponent. Body mass can be transferred into an attack from high to low or from low to high.

Rapid Retraction

When Marines deliver a punch, rapid retraction of the fist is important. Once the hand has made contact with the target, Marines quickly return to the basic warrior stance. Rapid retraction—

- Returns the hand and arm to the protection afforded by the basic warrior stance.
- Prevents the opponent from being able to grab the hand or arm.
- Permits the hand and arm to be "chambered" or "re-cocked" in preparation for delivering a subsequent punch.

Telegraphing

Telegraphing a strike occurs when body movements inform the opponent of the intention to launch an attack. Staying relaxed helps to reduce telegraphing.

Often, an untrained fighter telegraphs his intention to attack by drawing his hand back in view of his opponent, changing facial expression, tensing neck muscles, or twitching. These movements, however small, immediately indicate an attack is about to be delivered. If the opponent is a trained fighter, he may be able to evade or counter the attack. If the opponent is an untrained fighter, he may still be able to minimize the effect of an attack.

2. Punches

Punches may be thrown during any hand-to-hand confrontation. Most people resort to punching because it is a natural reaction to a threat. The purpose of a punch is to stun the opponent or to set him up for a follow-up finishing technique. However, punches should only be executed to the soft tissue areas of an opponent. A correctly delivered punch maximizes the damage to an opponent while minimizing the risk of injury to Marines.

Basic Fist

Punches are executed using the basic fist. To make the basic fist, the fingers are curled naturally into the palm of the hand and the thumb is placed across the index and middle fingers. Do not clench the fist until movement has begun. This reduces muscular tension in the forearm and increases speed and reaction time. Just before impact, Marines exert muscular tension on the hand and forearm to maximize damage to the opponent and reduce their chances of injury. Contact should be made with the knuckles of the index and middle fingers.

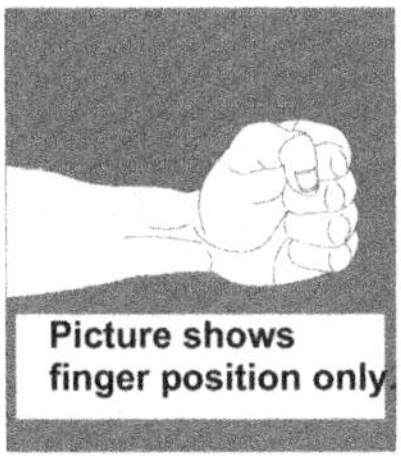

When striking with the basic fist, Marines must keep the hand straight, or in line, with the wrist to avoid injury to the wrist.

Lead Hand Punch

The lead hand punch is a snapping straight punch executed by the forward or lead hand. It is a fast punch designed to keep the opponent away and to set up an attack. A lead hand punch conceals movement and allows Marines to get close to the opponent. Lead hand punches should strike soft tissue areas, if possible. To execute the lead hand punch, Marines—

Snap the lead hand out to nearly full extension, while rotating the palm to the ground.

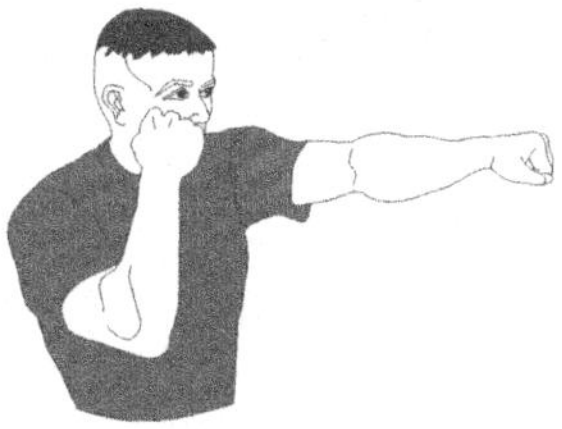

Contact the opponent with the first two knuckles of the fist.

Retract the hand immediately, resuming the basic warrior stance.

Rear Hand Punch

The rear hand punch is a snapping punch executed by the rear (right) hand. It is a power punch designed to inflict maximum damage on the opponent. Its power comes from pushing off the rear leg and rotating the hips and shoulders. To execute the rear hand punch, Marines—

Rotate the hips and shoulders forcefully toward the opponent and thrust the rear hand straight out, rotating the palm down, to nearly full extension.

Shift body weight to the lead foot while pushing off on the ball of the rear foot.

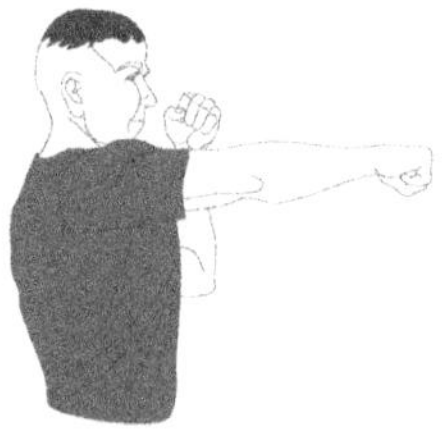

Contact the opponent with the first two knuckles of the fist.

Retract the hand immediately.

Uppercut

The uppercut is a powerful punch originating below the opponent's line of vision. It is executed in an upward motion traveling up the centerline of the opponent's body. It is delivered in close and usually follows a preparatory strike that leaves the target area unprotected. When delivered to the chin or jaw, the uppercut can render an opponent unconscious, cause extensive damage to the neck, or sever the tongue. To execute the uppercut, Marines—

Bend the arms, rotating the palm inboard. The distance the arms bend depends on how close the opponent is.

Rotate the hips and shoulders forcefully toward the opponent, thrusting the fist straight up toward the opponent's chin or jaw.

Contact the opponent with the first two knuckles of the fist.

Retract the hand immediately.

Hook

The hook is a powerful punch that is executed close in and is usually preceded by a preparatory strike. To execute the hook, Marines—

Thrust the right arm in a hooking motion toward the opponent, keeping the elbow bent while forcefully rotating the right shoulder and hip toward the opponent.

Contact the opponent with the first two knuckles of the fist. Continue rotating the shoulder and hip, following through with the fist to the target.

Retract the hand immediately.

3. Strikes with the Upper Body

Strikes stun the opponent or set him up for a follow-up finishing technique. The hands, forearms, and elbows are individual weapons of the arms that can be used to execute strikes including the hammer fist, chin jab, knife hand, eye gouge, and elbow strikes. These strikes provide a variety of techniques that can be used in any type of close combat situation.

Principles of Execution

There are several principles of execution that ensure a strike's effectiveness.

Generating Power. Weight transfer is necessary to generate maximum power in a punch. Marines accomplish this by—

- Rotating the hips and shoulders into the attack.
- Moving the body mass straight forward or backward in a straight line.
- Dropping body weight into an opponent. Body mass can be transferred into an attack from high to low or from low to high.

Muscular Tension. The arms are relaxed until the moment of impact. At the point of impact, Marines apply muscular tension in the hand and forearm to maximize damage to the opponent and to avoid injury to the hand. The arms are relaxed until the moment of impact.

Hit and Stick and Follow-Through. A strike should be delivered so that the weapon (e.g., hand, elbow) hits and remains on the impact site (target) and follows through the target. This technique inflicts maximum damage on the opponent.

Strikes with the arms are executed with "heavy hands"; i.e., the strike is executed by driving through with the strike to allow the weight of the hand to go through the target area of the body. Contact with the opponent should be made with the arm slightly bent, and the arm extends as it moves through the target. This technique allows Marines to deliver strikes effectively without executing full force.

Movement. Movement puts Marines in the proper position for launching an attack against an opponent as well as providing protection.

Movement is initiated from the basic warrior stance and ends by resuming the basic warrior stance. Strikes can be performed with either the left or right arm depending upon—

- The angle of attack.
- The position of the opponent.
- The opponent's available, vulnerable target areas.

Target Areas of the Body

For each strike, there are target areas of the body that, when struck, maximize damage to an opponent. Strikes use gross motor skills as opposed to fine motor skills. The target areas of the body are just that—areas. Pinpoint accuracy on a specific nerve is not needed for the strike to be effective.

Hammer Fist

Striking with the hammer fist concentrates power in a small part of the hand, which, when transferred to the target, can have a devastating effect. The striking surface of the hammer fist is the meaty portion of the hand below the little finger. To execute the hammer fist strike, Marines—

Make a fist and bend the arm at approximately a 45- to 90-degree angle. At the same time, rotate the right hip and right shoulder backward.

Thrust the fist forward onto the opponent while rotating the right hip and shoulder forward.

Rotate the wrist so the hammer fist makes contact on the opponent.

Chin Jab

The chin jab can immediately render an opponent unconscious and cause extensive damage to the neck and spine. The striking surface is the heel of the palm of the hand. To execute the chin jab, Marines—

Bend the right wrist back at a 90-degree angle with the palm facing the opponent and the fingers pointing up.

Keep the right arm bent and close to the body. Extend the hand into a concave position with the fingers slightly spread apart.

Step forward with the left foot toward the opponent, keeping the feet approximately shoulder-width apart and the knees bent. This is done to close with the opponent.

Keep the right arm bent and close to the side. Thrust the palm of the hand directly up under the opponent's chin. At the same time, rotate the right hip forward to drive the body weight into the attack to increase the power of the strike. The attack should travel up the centerline of the opponent's chest to his chin.

Knife Hand

The knife hand is one of the most versatile and devastating strikes. The striking surface is the cutting edge of the hand, which is the meaty portion of the hand below the little finger extending to the top of the wrist. The striking surface is narrow, allowing strikes on the neck between the opponent's body armor and helmet. The knife hand strike is executed from one of three angles: outside, inside, and vertical.

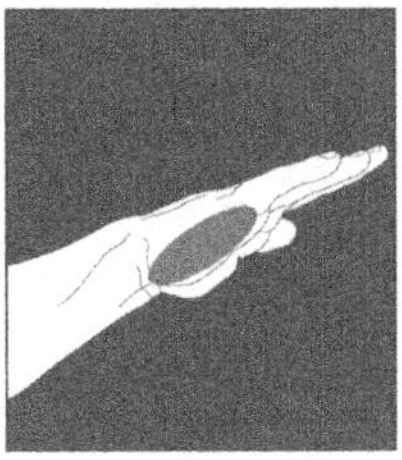

Outside Knife Hand. To execute the outside knife hand strike, Marines—

Execute a knife hand by extending and joining the fingers of the right hand and placing the thumb next to the forefinger (like saluting).

Retract the right hand. At the same time, rotate the right hip and right shoulder backward.

Thrust the knife hand forward (horizontally) onto the opponent while rotating the right hip and shoulder forward.

Inside Knife Hand. To execute the inside knife hand strike, Marines—

Execute a knife hand.

Bring the right hand over the left shoulder. At the same time, rotate the right shoulder forward and the left hip forward.

Thrust the knife hand forward (horizontally) onto the opponent while rotating the right hip and shoulder forward and the left shoulder backward.

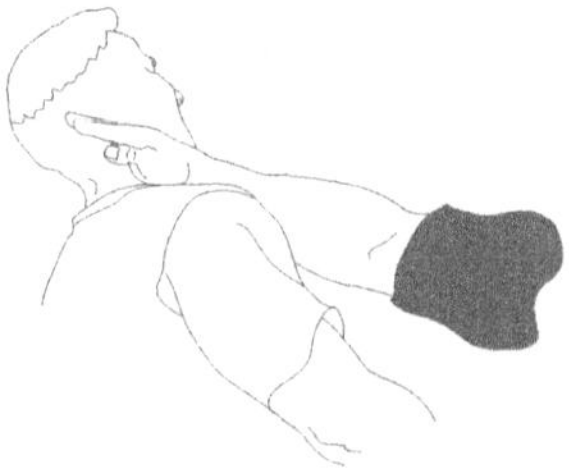

Vertical Knife Hand. When thrown vertically, the knife hand strike comes straight down in a straight line.

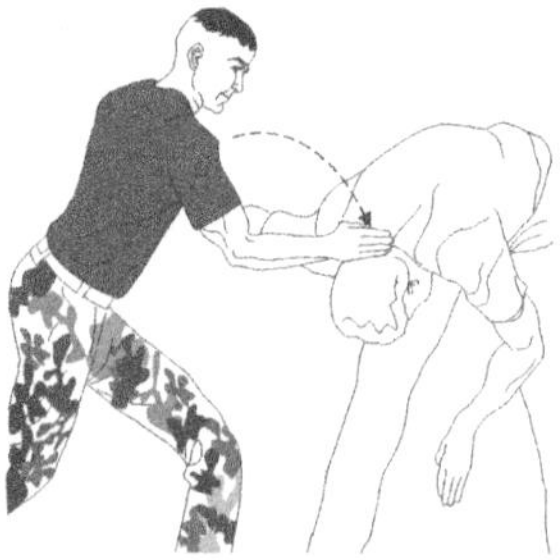

Eye Gouge

The eye gouge is used to attack an opponent's eyes, blinding him so follow-up strikes can be executed. The striking surface is the tips of the fingers or thumb. To execute the eye gouge, Marines—

Extend the right hand with the fingers slightly spread apart to allow entry into the eye sockets.

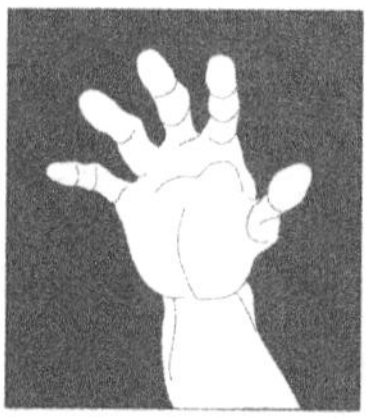

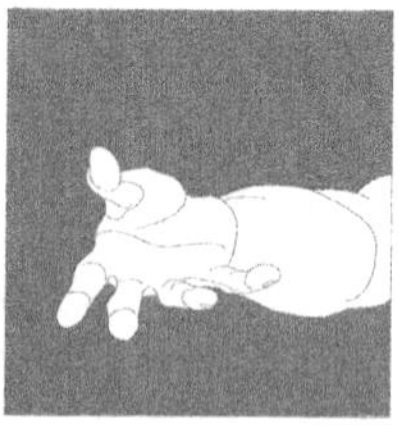

Place the palm of the hand either toward the ground or toward the sky and thrust the right hand forward into the opponent's eyes.

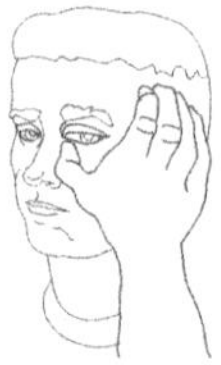

Thrust the hand forward at the opponent's nose level so the fingers or thumb slide naturally into the grooves of the opponent's eye sockets.

Elbow Strikes

The elbow is a powerful weapon that can be used in several different ways to attack virtually any part of an opponent's body. Elbow strikes can be performed either vertically (upward or downward) or horizontally (forward or reverse). The striking surface is 2 inches above or below the point of the elbow, depending upon the angle of attack, the opponent's attack angle, and the position of the opponent.

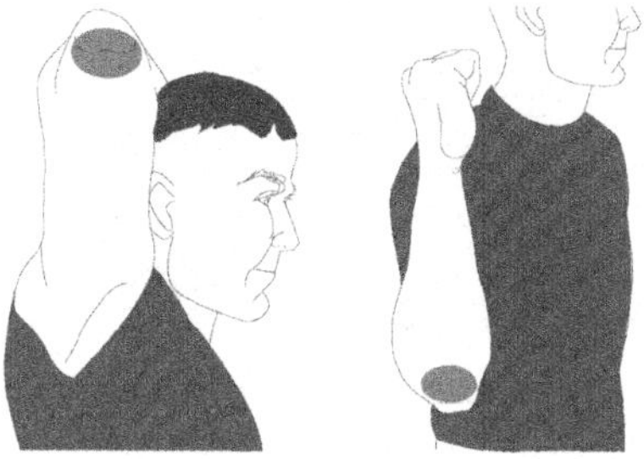

Above Elbow **Below Elbow**

Vertical Elbow Strike (Up). To execute an upward vertical elbow strike, Marines—

Bend the right elbow, keeping the fist close to the body. The fist is at shoulder level and the elbow is next to the torso.

Thrust the elbow vertically upward toward the opponent while rotating the right shoulder and hip forward to generate additional power.

Contact the opponent with the right forearm 2 inches above the point of the elbow.

Vertical Elbow Strike (Down). To execute a downward vertical elbow strike, Marines—

Bend the right elbow, keeping the fist close to the body. The fist is on the shoulder and the elbow is raised well above the shoulder.

Thrust the elbow vertically downward toward the opponent while dropping body weight into the attack to generate additional power.

Contact the opponent with the right triceps 2 inches above the point of the elbow.

Horizontal Elbow Strike (Forward). To execute the forward horizontal elbow strike, Marines—

Tuck the right fist near the chest with the palm heel facing the ground.

Thrust the right elbow horizontally forward toward the opponent. The forearm is parallel to the ground.

Rotate the right shoulder and hip forward to generate additional power.

Contact the opponent with the right forearm 2 inches below the point of the elbow.

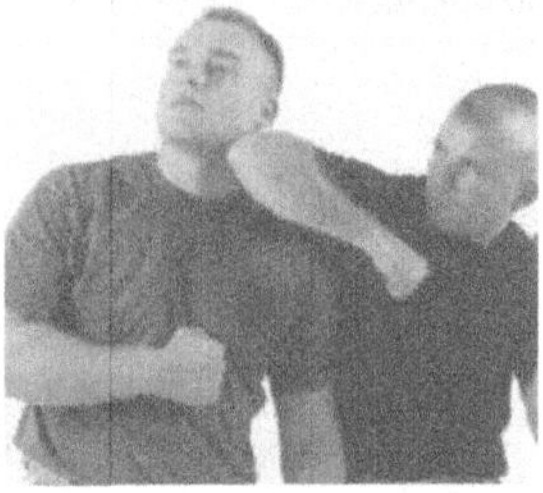

Horizontal Elbow Strike (Rear). To execute the rear horizontal elbow strike, Marines—

Tuck the right fist near the left shoulder with the palm heel facing the ground. At the same time, rotate the right shoulder forward and the left hip forward.

Thrust the right elbow horizontally rearward toward the opponent. The forearm is parallel to the ground and the hand moves toward the direction of the attack.

Rotate the right hip back and the right shoulder backward to generate additional power.

Contact the opponent with the right triceps 2 inches above the point of the elbow.

4. Strikes with the Lower Body

The legs are the body's most powerful weapons because they use the largest muscle groups to generate a strike. Legs are also less prone to injury. The feet are the preferred choice for striking because they are protected by boots. Marines use their feet, heels, and knees to execute kicks, knee strikes, and stomps.

Kicks

The purpose of a kick is to stop an opponent's attack or to create an opening in his defense in order to launch an attack. Kicks can be performed with the left (lead) leg or the right (rear) leg. Kicks with the rear leg have greater power because the hips are rotated into the attack. However, the rear leg is further away from the opponent so a strike with the rear leg will not contact the opponent as quickly as a strike with the lead leg.

Front Kick. The front kick is executed when the opponent is in front of the Marine. The front kick, delivered with either the rear or lead leg, is effective for striking below the waist. Attempting to kick higher results in diminished balance and provides the enemy with a greater opportunity to grab the leg or foot. The striking surfaces are the toe of the boot or the bootlaces. To execute the front kick, Marines—

Raise the left knee waist high, pivot the hips into the attack, and thrust the left foot forward toward the opponent.

Contact the opponent with the toe of the left boot or bootlaces.

Return to the basic warrior stance.

Side Kick. The side kick, delivered with the lead leg, is effective for striking the knees. The side kick is executed when the opponent is to the side of the Marine. The striking surface is the outside cutting edge of the boot near the heel. To execute the side kick, Marines—

Raise the right knee waist high and rotate the right hip forward.

Thrust the right foot to the right side toward the opponent, turning the foot at a 90-degree angle to maximize the striking surface on the opponent.

Contact the opponent with the cutting edge of the right boot.

Return to the basic warrior stance.

Knee Strikes

Knee strikes are excellent weapons during the close range of close combat fighting. The knee strike is generally delivered in close.

Vertical Knee Strike. The striking surface is the thigh, slightly above the knee. To execute the vertical knee strike, Marines—

Raise the right knee and drive it up forcefully into the opponent. Power is generated by thrusting the leg upward.

Contact the opponent 2 inches above the right knee.

Horizontal Knee Strike. The horizontal knee strike is executed with the leg generally parallel with the ground while rotating the hips to generate power. It is often delivered to the peroneal nerve. The striking surface is the front of the leg, slightly above or below the knee. To execute the horizontal knee strike, Marines—

Raise the right knee, rotate the right hip forward while pivoting on the left foot, and drive the knee horizontally into the opponent.

Contact the opponent 2 inches above the right knee.

Stomps

Stomps are delivered with the feet, usually when the opponent is down. Remember, when the opponent is down, Marines take whatever target is available.

Vertical Stomp. The vertical stomp allows Marines to remain upright and balanced, to rapidly deliver multiple blows with either foot, and to quickly and accurately attack the target. It is the preferred stomp. The striking surface is the flat bottom of the boot or the cutting edge of the heel. To execute the vertical stomp, Marines—

Raise the right knee above the waist with the right leg bent at approximately a 90-degree angle.

Drive the flat bottom of the right boot or the cutting edge of the right heel down onto the opponent forcefully. At the same time, bend the left knee slightly to drop the body weight into the strike.

Ax Stomp. The striking surface of the ax stomp is the cutting edge of the heel. To execute the ax stomp, Marines—

Raise the right heel above the waist, keeping the right leg straight.

Drive the cutting edge of the right heel down onto the opponent forcefully. At the same time, bend the left knee slightly to drop the body weight into the strike.

5. Counters to Strikes

In a close combat situation, an opponent will attempt to strike Marines with punches and kicks. When an opponent uses a strike, Marines must first avoid the strike. This is accomplished by moving quickly and blocking. Next, Marines must get into an offensive position. This allows Marines to use offensive strikes to attack the opponent. Regardless of the strike, the counter to a strike requires Marines to move, block, and strike.

Move

The first step in countering a strike is to move out of the way of the strike's impact. Movement removes Marines from the opponent's intended strike point and positions Marines to attack. Movement is executed at approximately a 45-degree angle to the front or rear. Movement is always initiated from the basic warrior stance. Return to the basic warrior stance with the toe of the lead foot pointing toward the opponent once the movement is complete.

Block

Different blocks are executed based on the strike. These will be covered by individual counters.

Strike

Any of the upper body or lower body strikes can be executed as a follow-on attack or part of the counter to an opponent's strike. The follow-on strike is determined by the angle to the opponent, the position of the opponent, and the opponent's available vulnerable target areas.

Counters to Punches

Counter to a Lead Hand Punch. This counter is used when the opponent throws a lead hand punch. To execute the counter to the lead hand punch, Marines—

Step forward and to the left at approximately a 45-degree angle, moving in to the outside of the opponent's attacking arm.

Raise the left arm and block or deflect the opponent's lead hand with the palm of the hand or the meaty portion of the forearm.

Hit and stick by leaving the left arm against the opponent's right arm while stepping forward and to the left at approximately a 45-degree angle to close with the opponent.

Execute a finishing technique, such as a strike or a kick, to the opponent's exposed target areas.

Counter to a Rear Hand Punch. This counter is used when the opponent throws a rear hand punch. To execute the counter to the rear hand punch, Marines—

Step forward and to the left at approximately a 45-degree angle, moving in to the outside of the opponent's attacking arm.

Raise the left arm and block or deflect the opponent's rear hand with the palm of the hand or the meaty portion of the forearm.

Hit and stick by leaving the left arm against the opponent's right arm while stepping forward and to the right at approximately a 45-degree angle to close with the opponent.

Execute a finishing technique, such as a strike or a kick, to the opponent's exposed target areas.

Counters to Kicks

Counter to a Front Kick (Left or Lead Leg). This counter is used when the opponent executes a front kick with his left leg. To execute the counter to a front kick, Marines—

Step forward and to the right at approximately a 45-degree angle, moving in to the outside of the opponent's striking leg.

Lower the left arm and block or deflect the opponent's leg with the palm of the hand or the meaty portion of the forearm.

Hit and stick by leaving the left arm against the opponent's leg while stepping forward and to the left at approximately a 45-degree angle to close with the opponent.

Execute a finishing technique, such as a strike or a kick, to the opponent's exposed target areas.

Counter to a Front Kick (Right or Rear Leg). This counter is used when the opponent executes a front kick with his right leg. To execute the counter to a front kick, Marines—

Step forward and to the left at approximately a 45-degree angle, moving in to the outside of the opponent's striking leg.

Lower the left arm and block or deflect the opponent's leg with the palm of the hand or the meaty portion of the forearm.

Hit and stick by leaving the left arm against the opponent's leg while stepping forward and to the right at approximately a 45-degree angle to close with the opponent.

CHAPTER 5

THROWS

This chapter describes all techniques for a right-handed person. However, all techniques can be executed from either side.

The Marine is depicted in camouflage utilities. The opponent is depicted without camouflage.

Marines use throwing techniques to maintain the tactical advantage and to throw the opponent to the ground during close combat. Throws apply the principles of balance, leverage, timing, and body position to upset an opponent's balance and to gain control by forcing the opponent to the ground. Throwing techniques are effective because they are size- and gender-neutral, and they rely on the momentum and power generated by the opponent rather than the strength or size of the Marine. Marines also execute a throw as a devastating attack against an opponent, possibly causing unconsciousness or broken limbs. When Marines execute throws, they must maintain balance and, simultaneously, prevent the opponent from countering a throw or escaping after being forced to the ground.

1. Turning Throw

Marines use a turning throw to take the opponent to the ground while they remain standing. A turning throw can also be executed from a stationary position. It is particularly effective if Marines and the opponent are wearing gear. To execute the turning throw, Marines—

Grasp the opponent's right wrist with the left hand.

Step forward with the right foot, place it against the outside of the opponent's right foot, and pivot so the back of the heel is next to the middle of the opponent's foot.

Hook the opponent's right arm with the right arm and pinch his arm between the biceps and forearm, touching the opponent with the body.

Pull the opponent's wrist downward, keeping it close to the body.

Pivot to the left on the ball of the foot and continue pulling downward on the opponent's wrist while rotating the wrist outward to off-balance him.

2. Hip Throw

Marines use a hip throw to take the opponent to the ground while they remain standing. A hip throw is particularly effective if the opponent is moving forward or pushing. Marines use the opponent's forward momentum to execute the hip throw. To execute the hip throw, Marines—

Grasp the opponent's right wrist with the left hand.

Step forward with the right foot and place it against the outside of the opponent's right foot.

The back of the heel should be next to the opponent's foot.

Step back with the left foot and rotate on the ball of the foot. The back of the heel is next to the opponent's toe. The knees are bent.

Rotate at the waist and hook the right hand around the back of the opponent's body (anywhere from his waist to his head). The side and hip should be against the opponent.

Rotate the hip against the opponent. The hips must be lower than the opponent's.

Use the right hand to pull the opponent up on the hip to maximize contact.

Pull the opponent's arm across the body and, at the same time, lift the opponent off the ground slightly by bending at the waist, straightening the legs, and rotating the body to the left.

3. Leg Sweep

Marines use a leg sweep to take the opponent to the ground while they remain standing. A leg sweep is particularly effective if the opponent is already off-balanced and moving backward or pulling on the Marine. To execute the leg sweep, Marines—

Grab the opponent's right wrist with the left hand and grab the opponent's left shoulder with the right hand.

Note: Grab the opponent's clothing or gear if his wrist and shoulder cannot be grabbed.

Step forward with the left foot and place it on the outside of the opponent's right foot.

Pull the opponent's wrist downward, close to the body, and push his shoulder backward to off-balance him.

Raise the right knee no higher than the waist.

Kick the foot past the opponent's right leg.

Use the heel of the boot to make contact with the opponent's calf (anywhere from the top of the calf down to the Achilles tendon or on the inside of the calf).

Sweep the opponent to the ground.

CHAPTER 6

CHOKES AND HOLDS

This chapter describes all techniques for a right-handed person. However, all techniques can be executed from either side.

The Marine is depicted in camouflage utilities. The opponent is depicted without camouflage.

WARNING

During training, never execute a choke at full force or full speed and never hold a choke for more than 5 seconds.

When Marines correctly perform a choke, they render an opponent unconscious in as little as 8 to 13 seconds. Chokes are easily performed regardless of size or gender. Marines must know how to apply chokes and how to counter a choke or a hold executed by an opponent.

1. Types of Chokes

There are two types of chokes: an air choke and a blood choke. An air choke closes off the airway to the lungs, thereby preventing oxygen from reaching the heart. A blood choke cuts off the blood flow to the brain. Both types can result in unconsciousness and eventual death for an opponent.

Air Choke

An air choke is performed on the opponent's windpipe (or trachea), cutting off the air to the lungs and heart. If Marines execute the air choke properly, the opponent loses consciousness within 2 to 3 minutes. Due to the length of time it takes to immobilize the opponent, air chokes are not recommended.

Blood Choke

A blood choke is performed on the opponent's carotid artery, which carries oxygen-enriched blood from the heart to the brain. The carotid artery is located on both sides of the neck. If Marines execute a blood choke properly, the opponent will lose consciousness within 8 to 13 seconds. The blood choke is the preferred choke because its intended effect (i.e., the opponent losing consciousness) can be executed quickly, ending the fight.

2. Chokes

Front Choke

Marines execute a front choke when they are facing the opponent. Marines use the opponent's lapels or collar to execute a front choke. To execute the front choke, Marines—

Grab the opponent's right lapel with the right hand, making certain that the knuckles or the back of the hand are against the opponent's carotid artery on the right side of his neck.

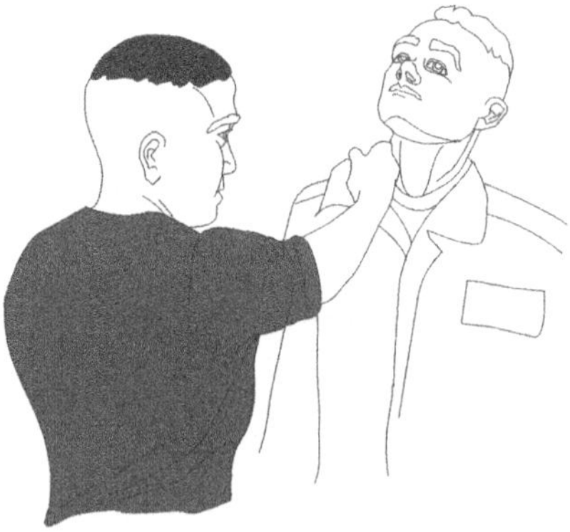

Keep the right hand against the opponent's neck, reach under the opponent's right arm with the left hand, grab the opponent's left lapel, and form an X with the wrists.

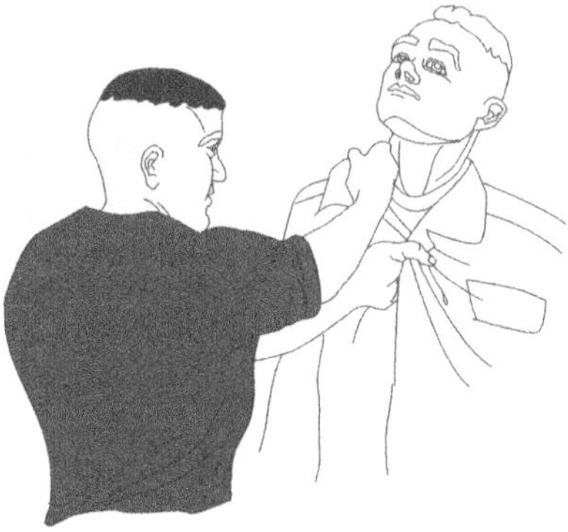

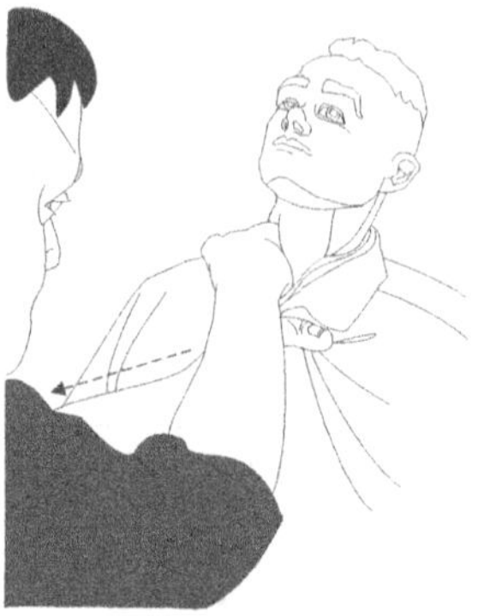

Press the right hand against the opponent's carotid artery.

Pull his lapel to the left with the left hand.

Side Choke

Marines execute a side choke when they are facing the opponent. The side choke is particularly effective when deflecting a punch thrown by an opponent. To execute the side choke, Marines—

Use the left hand to parry the opponent's arm inboard (to the inside of the opponent's reach).

Bring the right arm underneath the opponent's arm and up around the front of his neck.

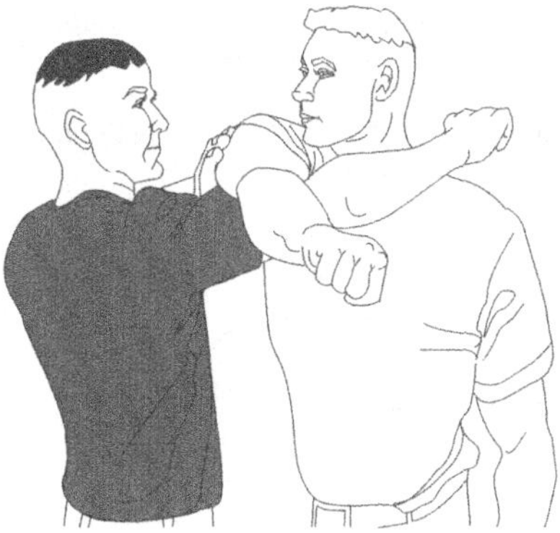

Extend the fingers, place the back of the forearm against the opponent's neck just below his ear, and press the carotid artery on his neck.

Note: If Marines are unable to place their fingers against the opponent's neck, they may use the back of the thumb or wrist.

Reach, with the left hand, around the back of the opponent's neck and clasp the hands together.

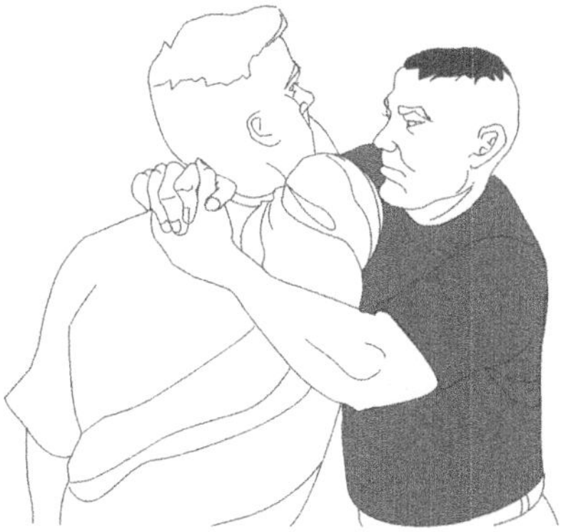

Pull the opponent toward the chest by pulling the clasped hands toward the chest.

Exert pressure on the side of the opponent's neck with the forearm.

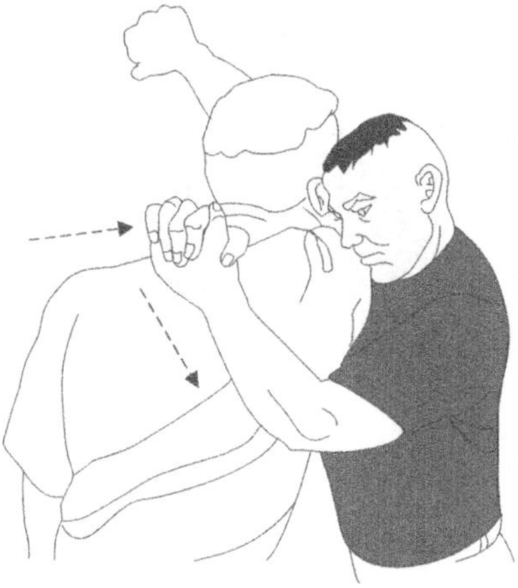

Rear Choke

Marines execute a rear choke when they are behind the opponent, the opponent is on the ground, or when they are taking the opponent to the ground. To execute the rear choke, Marines—

Reach, with the right arm, over the opponent's right shoulder and hook the bend of the arm around his neck.

Clasp both hands together.

Exert pressure with the biceps and forearms on both sides of the opponent's neck on his carotid arteries.

Maintain pressure with the biceps and forearms on both sides of the neck and draw the opponent closer by drawing the right arm in.

Figure-Four Choke

A variation of the rear choke is the figure-four choke. The figure-four choke allows Marines to gain more leverage than the rear choke. If the rear choke cannot be secured, the figure-four variation is employed to increase the pressure of the choke on the opponent. To execute the figure-four variation of the rear choke, Marines—

Apply a rear choke. The Marine's body should be against the opponent's body.

Grasp the left biceps with the right hand and place the left hand against the back of the opponent's head.

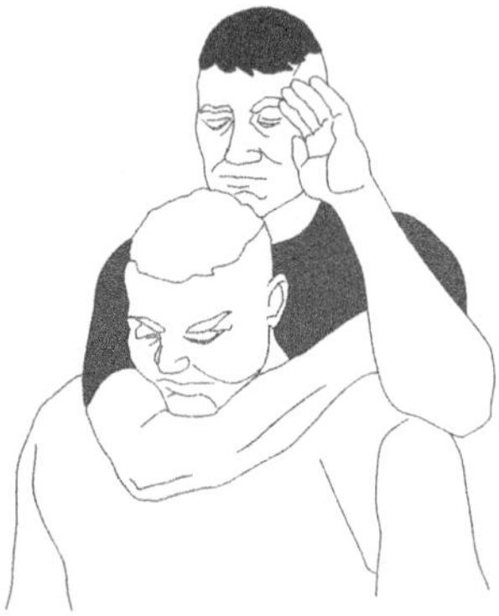

Push the opponent's head forward and down with the left hand.

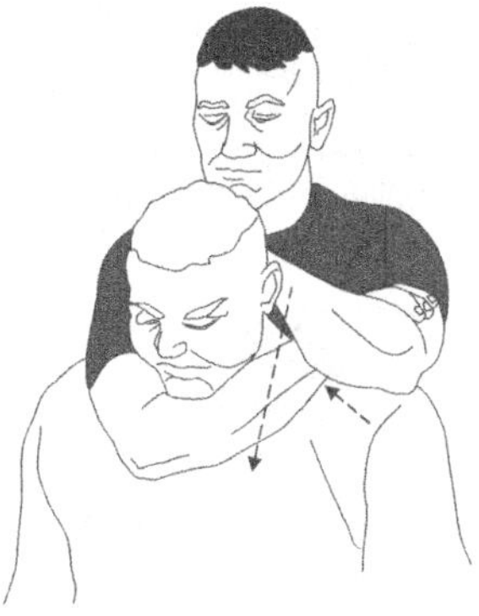

Draw the right arm in, maintaining pressure with the bicep and forearm on both sides of the opponent's neck.

3. Counters to Chokes and Holds

During a close combat situation, an opponent may apply a choke or hold on a Marine. If the opponent correctly applies a choke, a Marine quickly loses consciousness. If a choke is not executed properly, it often results in a hold, typically a bear hug or a headlock. A hold allows the opponent to control a Marine and removes the Marine's ability to attack. It is important for Marines to extract themselves from chokes and holds, regain the tactical advantage, and counter with strikes.

Although a choke causes unconsciousness in 8 to 13 seconds for a blood choke and 2 to 3 minutes for an air choke, the first movement in any counter to a choke is to clear the airway. Marines use softening techniques to loosen an opponent's grip and to clear their airway. Softening techniques are particularly effective if Marines lack the physical strength of their opponent. These techniques include groin strikes, eye gouges, foot stomps, etc. Softening techniques are not offensive; rather, they are used to loosen an opponent's hold.

The second movement is to tuck the chin. Once the airway is clear, Marines tuck their chins to prevent the opponent from reapplying the choke.

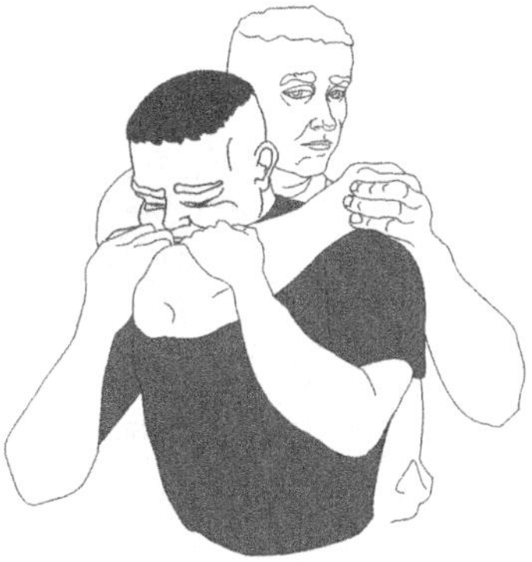

Counter to a Front Choke

Marines use a counter to a front choke when the opponent approaches from the front and uses both hands to choke a Marine around the throat. To execute the counter to the front choke, Marines—

Grasp the opponent's right forearm (where the elbow bends) with the left hand and apply downward pressure on the opponent's radial nerve with the fingers.

Execute a chin jab to the opponent's chin with the right hand. To generate power into the strike, bring the left foot to the outside of the opponent's right foot and rotate the hips into the strike.

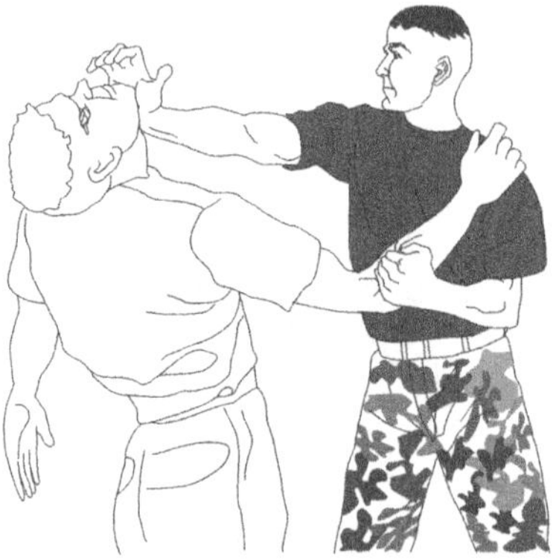

Counter to a Rear Choke

Marines execute a counter to a rear choke when the opponent approaches from the rear and puts his right arm around a Marine's throat. To execute the counter to the rear choke, Marines—

Grasp the opponent's forearm (at the radial nerve) and bicep with both hands and pull down just enough to clear the airway. Once the airway is clear, tuck the chin to protect the airway and to prevent the opponent from reapplying the choke.

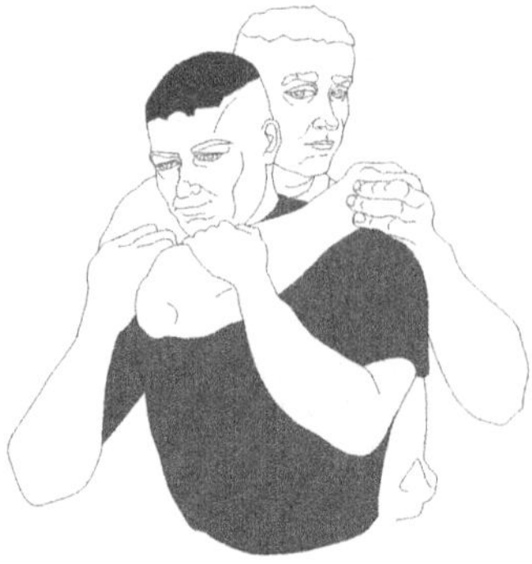

Step behind the opponent's right leg with the left foot, keeping both legs bent (almost in a squatting position).

Strike and drive the left elbow into the opponent's torso while rotating the hips and pivoting to the left, throwing the opponent back and to the ground.

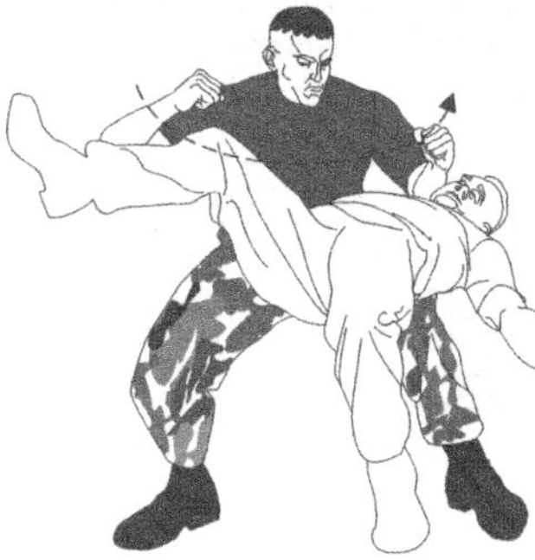

Counter to a Front Headlock

Marines use a counter to a front headlock when the opponent approaches from the front and puts his right arm around the Marine's neck, bends the Marine forward, and locks the Marine's head against his hip. To execute the counter to a front headlock, Marines—

Grasp the opponent's wrist and forearm with both hands and pull down to clear the airway. Maintain control of the opponent's wrist throughout the move. Once the airway is clear, tuck the chin to protect the airway and to prevent the opponent from reapplying the choke.

Move the right hand and arm across the opponent's torso.

Step forward and to the left with the left foot at a 45-degree angle.

Execute a sweep with the right foot against the opponent's right leg. At the same time, push against the opponent's chest with the right arm and shoulder to generate power in the sweep.

Counter to a Rear Headlock

Marines use a counter to a rear headlock when the opponent approaches from the rear and puts his right arm around the Marine's neck, bends the Marine forward, and locks the Marine's head against his hip. To execute the counter to the rear headlock, Marines—

Grasp the opponent's wrist and forearm with the right hand and pull down to clear the airway. Once the airway is clear, tuck the chin to protect the airway and to prevent the opponent from re-applying the choke.

Reach over the opponent's right shoulder with the left arm.

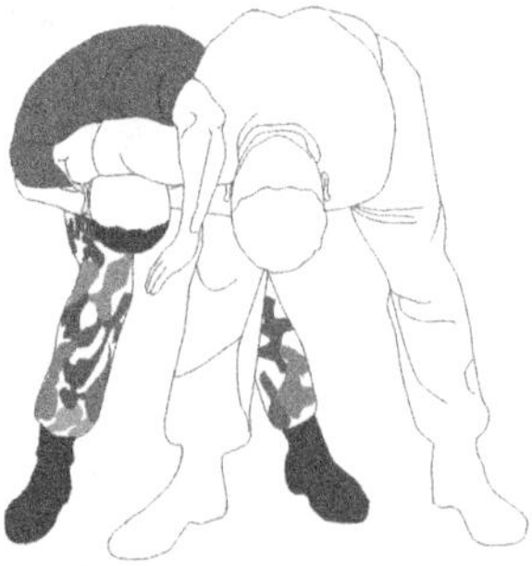

Grab any part of the opponent's face (chin, nose, eyes) and pull back while rising to a standing position.

Execute, with the right hand, a hammer fist strike to the opponent's exposed throat.

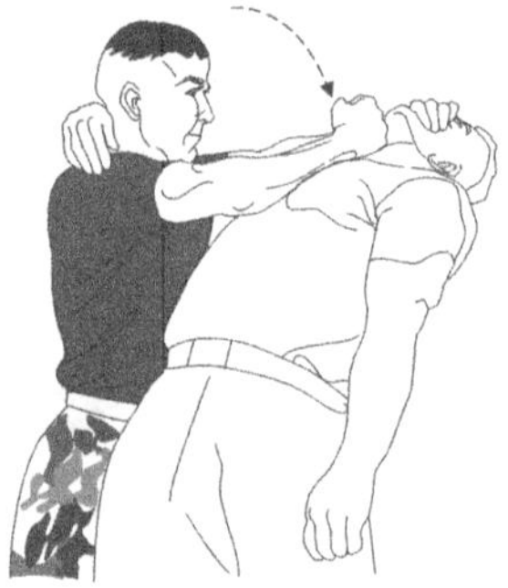

Counter to a Front Bear Hug

Marines execute a counter to a front bear hug when the opponent approaches from the front and puts both of his arms around the Marine's body, trapping the Marine's arms to the sides. To execute the counter to a front bear hug, Marines—

Step forward and to the left with the left foot at a 45-degree angle to the outside of the opponent's right leg, keeping the left leg bent.

Grasp the opponent's torso or arms to gain balance and to assist in throwing the opponent. It may be helpful to hook the opponent's right arm with the left arm.

Drive the right arm and shoulder forward and, at the same time, bring the right leg forward and sweep the opponent's right leg, bringing him to the ground.

Counter to a Rear Bear Hug

Marines use a counter to a rear bear hug when the opponent approaches from the rear and puts both of his arms around the Marine's body, trapping the Marine's arms to the sides. To execute the counter to a rear bear hug, Marines—

Step behind the opponent's right leg with the left foot, keeping both legs bent (almost in a squatting position). The left side of the body should be against the opponent's.

Pivot the hip, turning the body to the left and throwing the opponent back over the bent leg.

CHAPTER 7

GROUND FIGHTING

This chapter describes all techniques for a right-handed person. However, all techniques can be executed from either side.

In drawings, the Marine is depicted in woodland camouflage utilities; the opponent is depicted without camouflage. In photographs, the Marine is depicted in woodland camouflage utilities; the opponent is depicted in desert camouflage utilities.

Marines should avoid being on the ground during a close combat situation because the battlefield may be covered with debris and there is an increased risk of injury. However, many close combat situations involve fighting on the ground. The priority in a ground fight is for Marines to get back on their feet as quickly as possible. In any ground fighting scenario, Marines will usually end up in one of four positions with the opponent. The offensive positions, in which Marines have a tactical advantage, are the guard and mount. The defensive positions, which are used as counters when the opponent has the tactical advantage, are the counter to the guard and the counter to the mount. Marines can also employ chokes during ground fighting to quickly end a fight.

1. Offensive Ground Fighting

Guard Position

Marines execute a guard position when the opponent is on top and the legs are wrapped around the opponent's legs. If the opponent is on top, he may try to choke the Marine, but the Marine still has the tactical advantage because the Marine is in a position to control the situation. To execute the guard position technique, Marines—

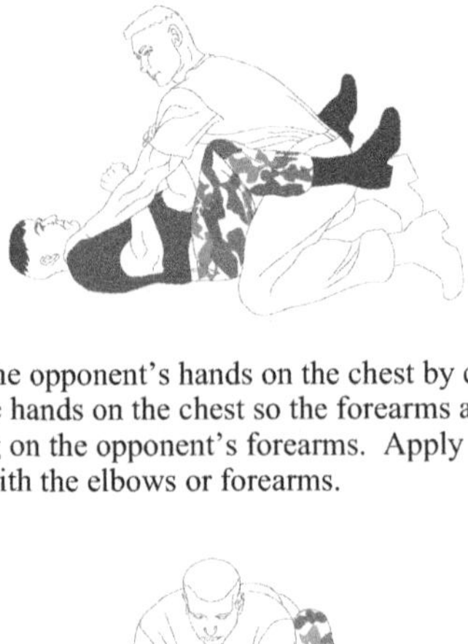

Trap the opponent's hands on the chest by crossing the hands on the chest so the forearms are resting on the opponent's forearms. Apply pressure with the elbows or forearms.

Strike the outside of the opponent's thigh with the cutting edge of the right heel. This causes the opponent to jerk to that side.

Move the head quickly to the left and swivel the hips to the right. At the same time, bring up both of the legs. Both legs are on the right side of the opponent's body.

Bring the right leg down, hooking the opponent's neck and head. Exert downward pressure to roll him over on his back. Grasp and maintain control of the opponent's left arm. Upon completion of the move, the Marine is sitting up with legs bent over the opponent while maintaining control of his arm.

Keep legs and knees bent. Maintain pressure against the opponent's neck with the back of the right foot and against his side with the left foot underneath his armpit. Squeeze the knees together, locking the opponent's arm.

Pull the opponent's arm straight up and fall back sharply, pulling his arm to the side in the direction of his little finger.

Return to standing.

Mount Position

Marines execute the mount if the opponent is lying on his back on the ground and the Marine is on top with legs wrapped around the opponent's body. This position is an offensive position because Marines are in a better position to control the opponent and to execute ground fighting techniques. To execute the mount position technique, Marines—

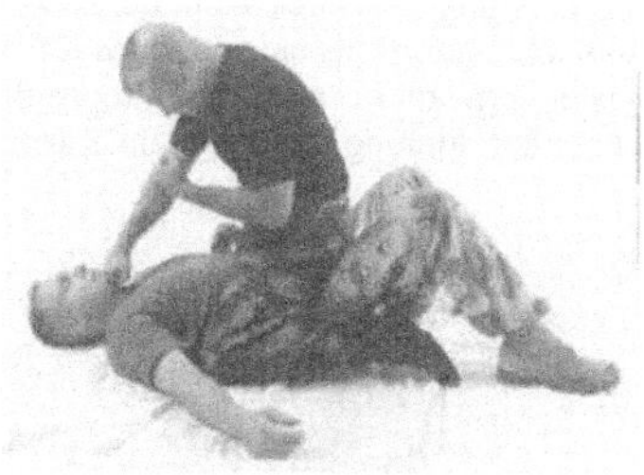

Grab the opponent's wrists or forearms with the hands. Hold them tightly against the torso.

Maintain control of the opponent's arms with the left arm and apply pressure to the opponent's brachial plexus (tie in) with the right hand.

Bring the left foot close to the opponent's right armpit and swing the right leg across the opponent's head.

Keep the legs and knees bent. Maintain pressure against the opponent's neck with the back of the right foot and against his side with the left foot which is underneath his armpit. Squeeze the knees together, locking the opponent's arm.

Pull the opponent's arm straight up and fall back sharply, pulling his arm to the side in the direction of his little finger.

Return to standing.

2. Defensive Ground Fighting

Counter to the Guard

Marines use this technique if the opponent is lying on his back on the ground and the Marine is kneeling on the ground between the opponent's legs. To execute the counter to the guard position, Marines—

Strike the opponent's femoral nerve, located on the inside of the thigh, with the elbows. This forces the opponent to separate his legs.

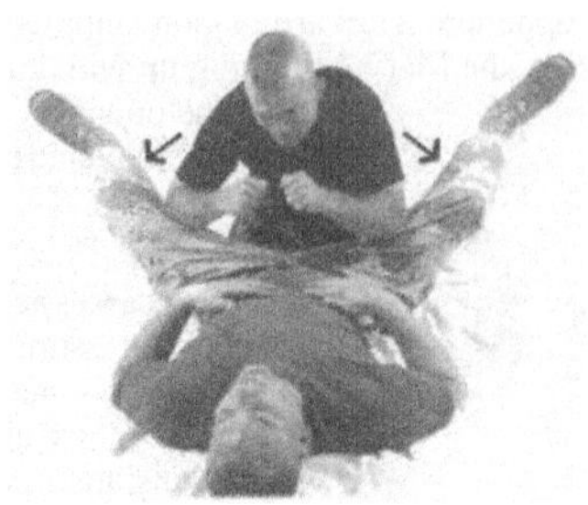

Strike the opponent's groin with the right fist.

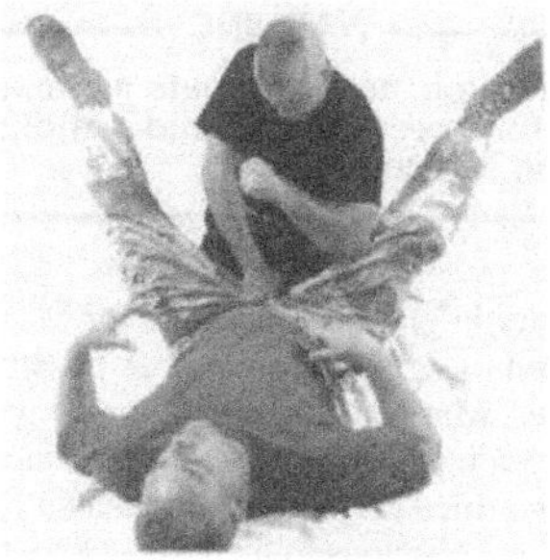

Hook the left arm underneath the opponent's right knee from the inside. Throw the opponent's leg over the head.

Duck the head quickly and move to the left.

Return to standing.

Counter to the Mount

Marines use this technique when they are lying on their back on the ground, the opponent is mounted on top of the Marine, and the opponent's legs are wrapped around the Marine. The opponent has the tactical advantage. To execute the counter to the mount position, Marines—

Grab the opponent's gear or clothing on his upper torso and pull him down close.

Use the right arm to hook the opponent's left arm, from the inside around the outside, above his elbow. With the right foot, hook the opponent's left leg or ankle.

Draw the elbow in to bend the opponent's elbow, bringing him down close. The arm must be hooked above the opponent's elbow in order to bend it.

Strike the opponent's side with the left hand.

Push the opponent over and roll him off to the right side.

Maintain control of the opponent's hooked arm and move to a kneeling position.

Move to a standing position while maintaining pressure on the opponent's arm and using the knee to apply pressure against the opponent's elbow.

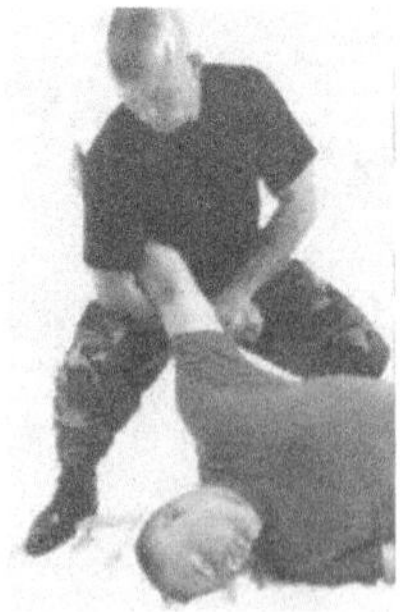

3. Ground Fighting Chokes

WARNING

During training, never execute a choke at full force or full speed. Never hold a choke for more than 5 seconds.

The priority in ground fighting is for Marines to get back on their feet as quickly as possible. Sometimes, Marines can quickly end a ground fight by executing a choke on the opponent. When performed correctly, a choke can render an opponent unconscious in as little as 8 to 13 seconds. Chokes are easily performed regardless of size or gender. The chokes performed during ground fighting are the same as those performed while standing (see chap. 6).

Ground Fighting Front Choke

The ground fighting front choke is a blood choke performed most effectively from the mount position. The front choke employs the opponent's lapels or collar to execute the choke. To execute the ground fighting front choke, Marines—

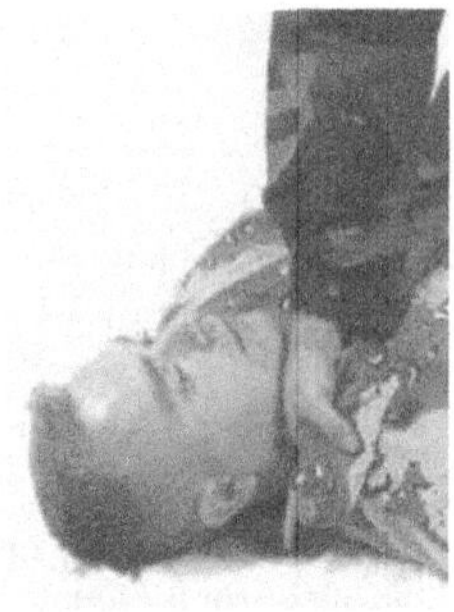

Grab the opponent's right lapel with the right hand, making certain that the knuckles or the back of the hand are against the carotid artery on the right side of the opponent's neck.

Keep the right hand pressed against the opponent's carotid artery, reach under the right arm with the left hand, and grab the opponent's left lapel, forming an X with the wrists.

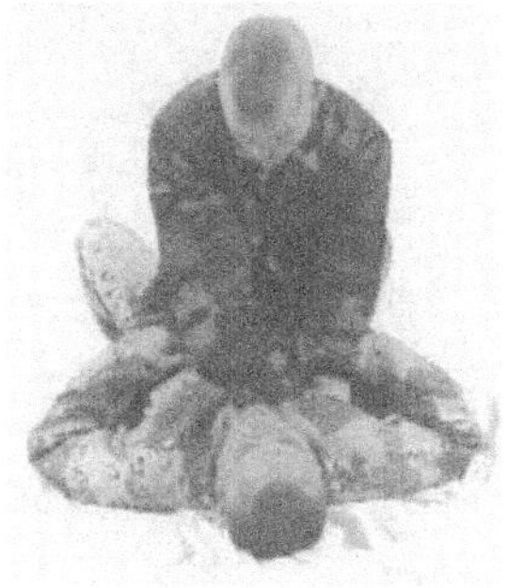

Keep the right hand pressed against the opponent's carotid artery and pull the opponent's left lapel to the left with the left hand.

Ground Fighting Side Choke

The ground fighting side choke is a blood choke performed from the mount position. The ground fighting side choke is particularly effective when the opponent raises his arms and places them on the Marine's chest or throat. To execute the ground fighting side choke, Marines—

Use the left hand to parry the opponent's right arm inboard (to the inside of the opponent's reach). Bring the right arm underneath the opponent's arm and up around the front of his neck.

Keep the fingers extended, place the back of the forearm against the opponent's neck just below his ear and press the carotid artery on his neck. Depending on the Marine's or the opponent's body configuration, the back of the thumb or wrist may press against the opponent's neck.

Use the left hand to reach around the back of the opponent's neck and clasp the hands together.

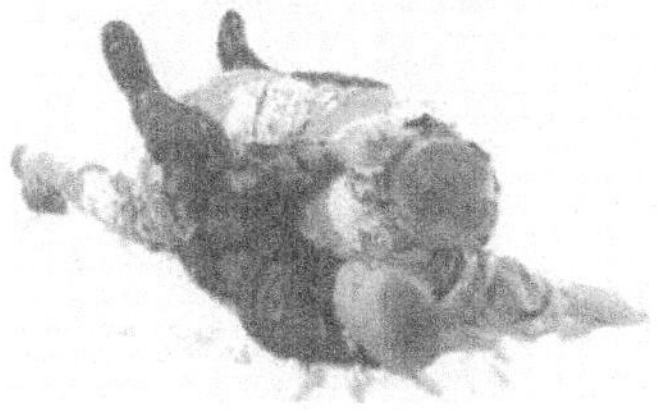

Pull the opponent toward the chest. Use the fore-arm to exert pressure on the side of his neck. This is done by pulling the clasped hands toward the chest.

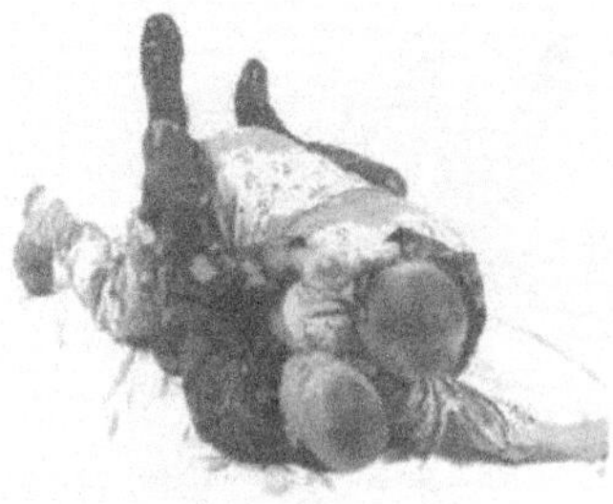

Ground Fighting Rear Choke

The ground fighting rear choke is a blood choke performed when Marines are behind the opponent. To execute the ground fighting rear choke, Marines—

Wrap the legs around the opponent with the left foot against the inside of his left thigh and the right foot against the inside of his right thigh.

Use the right arm to reach over the opponent's right shoulder and hook the bend of the arm around his neck.

Use the left hand to clasp both hands together.

Exert pressure with the biceps and forearms on both sides of the opponent's neck. This applies pressure to the opponent's carotid arteries. While maintaining pressure with the biceps and fore-arms, draw the opponent closer by drawing the right arm in. To increase the effectiveness of the choke, lean backward by arching the back and pulling the opponent back. At the same time, push the feet against the opponent's thighs.

Figure-Four Choke

A variation of the ground fighting rear choke is the figure-four choke. The figure-four choke allows Marines to gain more leverage on the rear choke. If the rear choke cannot be secured, the figure-four variation may be applied to increase the pressure of the choke on the opponent. To execute the figure-four variation of the rear choke, Marines—

Apply the ground fighting rear choke. The chest should be against the opponent's back.

Grasp the left biceps with the right hand and place the left hand against the back of the opponent's head.

Use the left hand to push the opponent's head forward and down.

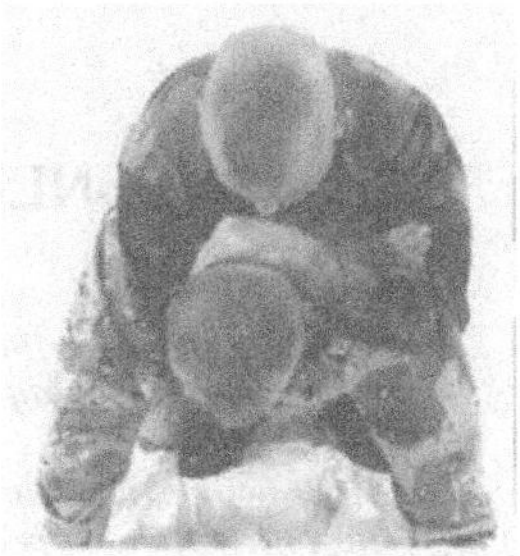

Draw the right arm in while maintaining pressure with the biceps and forearm on both sides of the opponent's neck.

Increase the effectiveness of the choke by arching backward and pushing the feet against the opponent's thighs. Continue to pull the opponent back with the right arm while exerting pressure forward with the left hand against the opponent's head.

CHAPTER 8

NONLETHAL TECHNIQUES

This chapter describes all techniques for a right-handed person. However, all techniques can be executed from either side.

In drawings, the Marine is depicted in woodland camouflage utilities; the opponent is depicted without camouflage. In photographs, the Marine is depicted in woodland camouflage utilities; the opponent is depicted in desert camouflage utilities.

The Marine Corps' involvement in military operations other than war—e.g., humanitarian, peacekeeping, or evacuation missions—has greatly increased. These missions require skills that span the spectrum of conflict and support operations within a continuum of force. But the Corps' day-to-day existence also demands a responsible use of force. Nonlethal techniques are among the skills Marines use to apply a responsible use of force.

1. Unarmed Restraints and Manipulation

Marines operate within a continuum of force, particularly in support of peacekeeping or humanitarian types of missions. In these situations, Marines must act responsibly to handle a situation without resorting to deadly force. Unarmed restraints and manipulation techniques including joint manipulation, come-alongs, and takedowns can be used to control a subject without resorting to deadly force. Marines must train to become proficient in nonlethal techniques and to respond in a responsible manner. These techniques are referred to as compliance techniques, and they are applied in the third level in the continuum of force.

WARNING

During training, never apply the techniques for unarmed restraints and manipulation at full force or full speed. Use a slow and steady pressure to avoid injury.

Compliance Techniques

Compliance techniques are unarmed restraint and manipulation techniques used to physically force a subject or opponent to comply. Compliance can be achieved through the close combat techniques of—

- Pain compliance using joint manipulation and pressure points. (Pain compliance is the initiation of pain to get compliance on the part of the subject.)
- Come-along holds.

Principles of Joint Manipulation

Joint manipulation is used to initiate pain compliance and gain control of a subject. It involves the application of pressure on the joints (elbow, wrist, shoulder, knee, ankle, and fingers). Pressure is applied in two ways:

- In the direction in which the joint will not bend. For example, joints such as the knees

and elbows only bend in one direction and when pressure is applied in the opposite direction, pain compliance can be achieved.
- Beyond the point where the joint stops naturally in its range of movement (i.e., it no longer bends).

Since each joint has a breaking point, Marines should apply slow steady pressure only until pain compliance is reached. Continued pressure will break the joint and may escalate the violence of the situation.

Joint manipulation also uses the principle of off-balancing. A subject can be better controlled when he is knocked off balance.

Wristlocks

A wristlock is a joint manipulation that can be applied in a number of ways to achieve pain compliance. The wrist rotates in a number of directions and will bend in a single direction until its movement stops naturally. In a wristlock, pressure is exerted beyond that point by bending or twisting the joint. A wristlock is executed when an opponent tries to grab Marines or is successful in grabbing Marines or their equipment. A wristlock can also be performed by Marines if they wish to initiate control of an opponent.

Basic Wristlock. A basic wristlock is executed when Marines grab the opponent's left hand with the right hand. To execute the basic wristlock, Marines—

Use the right hand to grab the opponent's left hand by placing the thumb on the back of the opponent's hand so that the Marine's knuckles are facing to the left.

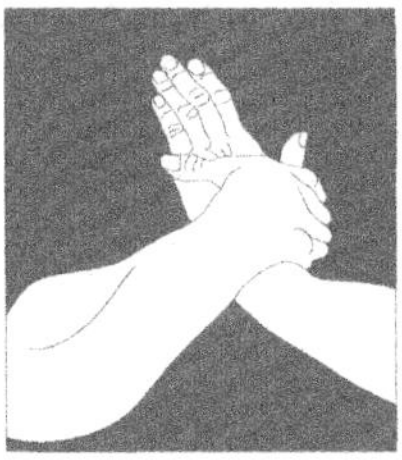

Hook the fingers across the fleshy part of the opponent's palm below the thumb. The fingers are used to anchor the hand so leverage can be applied to twist and bend the joint.

Exert downward pressure with the thumb to bend the opponent's joint. Rotate the opponent's hand to the right to twist the joint.

Use the left hand to further control the opponent.

Step in to the opponent to keep the opponent's hand in close to the body to control him and provide more leverage on the wristlock.

Note: When executing the basic wristlock with the left hand, the Marine grabs the opponent's hand so that the Marine's knuckles are facing to the right, and then rotates and twists the opponent's hand to the left.

Reverse Wristlock. A reverse wristlock is executed when Marines grab the opponent's right hand with the right hand. To execute the reverse wristlock, Marines—

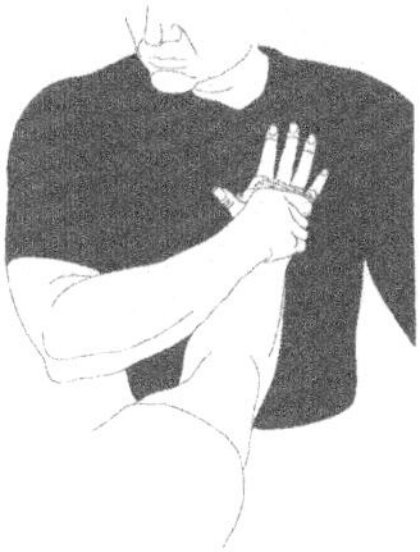

Place the right palm on the back of the opponent's right hand and wrap the fingers across the fleshy part of his palm below his little finger.

Twist the opponent's hand to the right while stepping in to place his hand against the chest. Apply downward pressure on the opponent's hand against the chest. Leave the opponent's hand on the chest to fully control the subject and to gain leverage.

Lean forward to use body weight to add additional pressure to the joint.

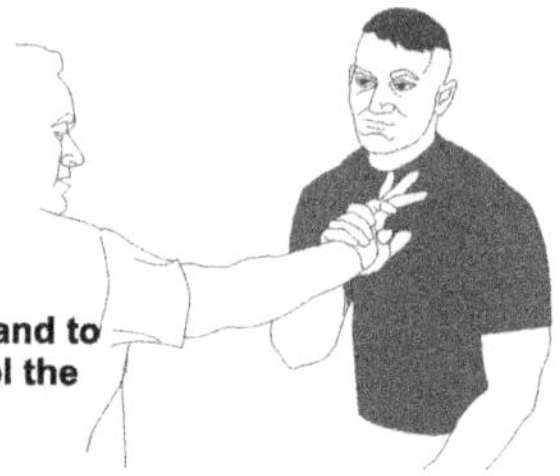

Use the left hand to further control the opponent.

Two-Handed Wristlock. Both hands can be used in the wristlock to maximize the leverage and pressure needed to bend and twist the joint. To execute the two-handed wristlock, Marines—

Place both thumbs on the back of the opponent's hand, thumbs crossed.

Hook the fingers of both hands around the fleshy part of the opponent's palm on both sides of his hand.

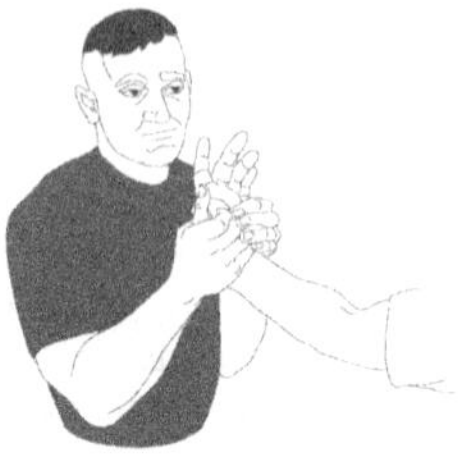

Step into the opponent and apply pressure downward on the back of his hand to bend the joint and rotate his wrist away from the body to twist the joint.

Enhanced Pain Compliance on Wristlock. Enhanced pain compliance techniques are applied in the third and fourth levels in the continuum of force. Additional pain be applied to a wristlock by—

Adding downward pressure to the elbow with the other hand or elbow by using the fingers to pull in on the opponent's radial nerve located on the inside of the forearm. When pressure is added to the opponent's radial nerve, his direction can be controlled.

Applying pressure against the opponent's finger joint to bend it in a direction it cannot bend (i.e., splitting the fingers).

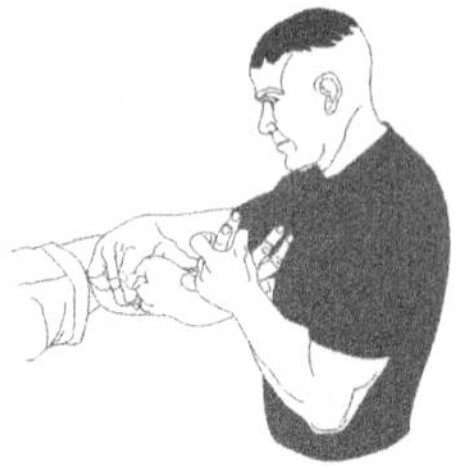

Come-Along Holds

Marines use a come-along hold to control and move an opponent.

Escort Position. A common come-along hold is the escort position. To execute the escort position, Marines—

Face the opponent. Use the left foot to step forward at a 45-degree angle. Turn to face the right side of the opponent.

Use the right hand to firmly grasp the opponent's right wrist. With the left hand, firmly grasp the opponent's right triceps.

Position the opponent's controlled arm diagonally across the torso, keeping his wrist against the right hip. The Marine should be standing to the right of and behind the opponent.

Note: This technique works well when escorting an opponent on either the right or left side. Take caution when escorting an opponent by ensuring his controlled hand is not in a position to grab the holstered weapon. The preferred escort position is from the Marine's left side, so that the opponent is kept further away from the weapon.

Wristlock Come-Along. To execute the wristlock come-along, Marines—

Use the left hand to execute a basic wristlock. Incorporate the right hand in a two-handed wristlock for more control.

Maintain pressure on the opponent's wrist with the right hand, step forward, and pivot around to stand next to the opponent.

Release the left hand, quickly reach under the opponent's arm from behind, and grab his hand.

Use the left hand and apply downward pressure on the opponent's wrist.

Controlling Technique. The following controlling technique is used when an opponent grabs the Marine's wrist. To execute the technique, Marines—

Trap the opponent's hand with the palm of the other hand.

Rotate the opponent's trapped hand up and on his forearm while maintaining downward pressure on his trapped hand.

Apply downward pressure with both hands until the opponent is taken to the ground.

Armbars

An armbar is a joint manipulation in which pressure is applied on a locked elbow, just above the joint, in the direction the joint will not bend. An armbar has to be locked in quickly, but still requires a slow, steady pressure to gain compliance.

Basic Armbar. To execute a basic armbar, Marines—

Use the right hand to grab the opponent's right wrist.

Bring the left hand down on or above the opponent's elbow joint. To gain additional leverage, pivot to face the opponent.

Use the left hand to apply downward pressure on or above the opponent's elbow joint while pulling up on his wrist.

Armbar from a Wristlock. To execute an armbar from a wristlock, Marines—

Use the right hand to grab the opponent's right hand and execute a reverse wristlock.

Bring the left hand down on or above the opponent's elbow joint.

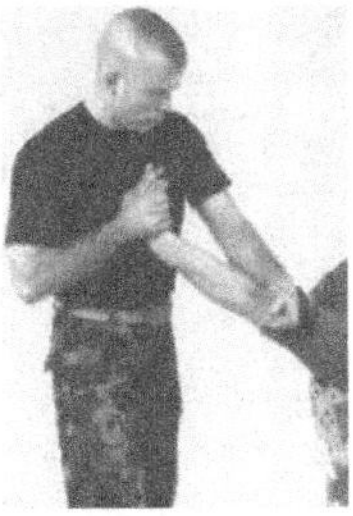

Use the left hand to apply downward pressure on or above the opponent's elbow joint while pulling up on his wrist.

Takedowns

A takedown is used to bring an opponent to the ground to further control him.

Takedown From a Wristlock Come-Along. To take the opponent to the ground from a wristlock come-along, Marines—

Use the right foot to push down on the opponent's calf or Achilles tendon.

Maintain control of the opponent's wrist and elbow and apply a slow, steady pressure to bring him to the ground.

Armbar to a Takedown. This technique is used to take a noncompliant opponent to the ground from an armbar. To execute the armbar takedown, Marines—

Use the right hand to execute a reverse wristlock.

Bring the left hand or forearm down on or above the opponent's elbow joint.

Pivot so the back is facing the opponent and, at the same time, lift the left elbow and slide the body so it is against the opponent, placing the armpit high above the opponent's elbow joint.

Lean back, placing the body weight on the opponent's arm until he complies or is taken to the ground.

Note: This technique may break the opponent's arm. Therefore, this technique should not be employed if the objective is a nonlethal takedown.

Wristlock Takedown. This technique is used to take a noncompliant opponent to the ground from a basic wristlock and to put him in a position where he can be handcuffed, if necessary. To execute the takedown, Marines—

Use the right hand to execute a basic wristlock. Incorporate the left hand in a two-handed wristlock.

Apply downward pressure on the wristlock, pivot on the ball of the right foot, and quickly turn to the right to take the opponent to the ground.

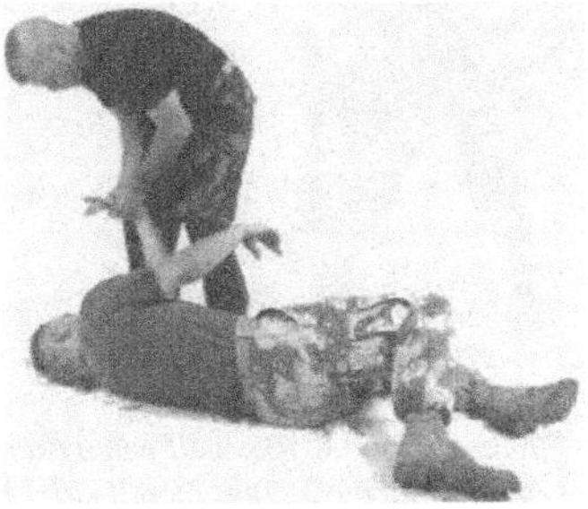

Continue to apply pressure on the wrist joint as the opponent lands on his back with his arm straight in the air.

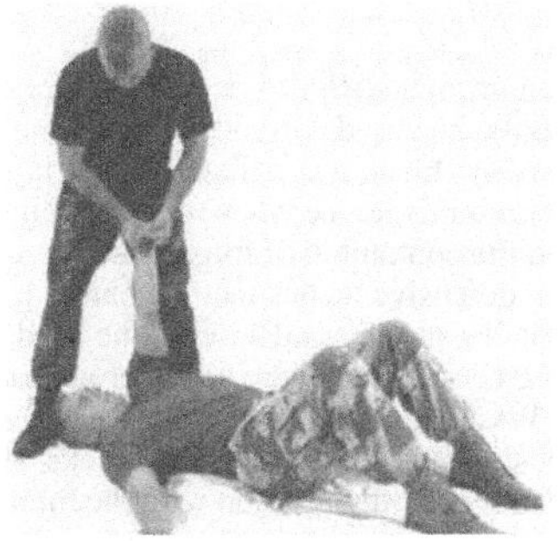

Slide the left foot under the opponent's back, underneath his armpit.

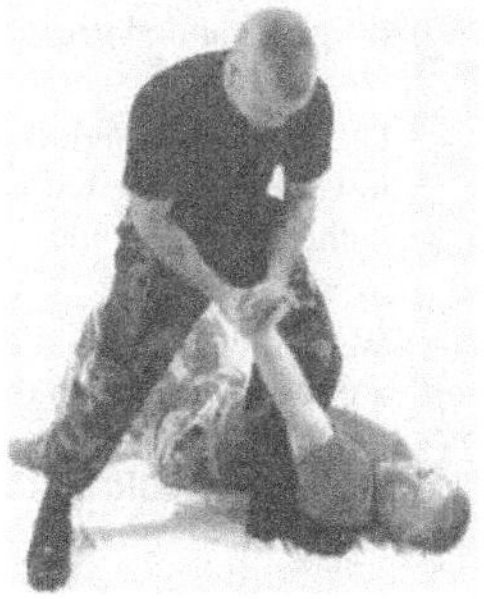

Apply pressure with the knee against the opponent's triceps while pulling back on his arm.

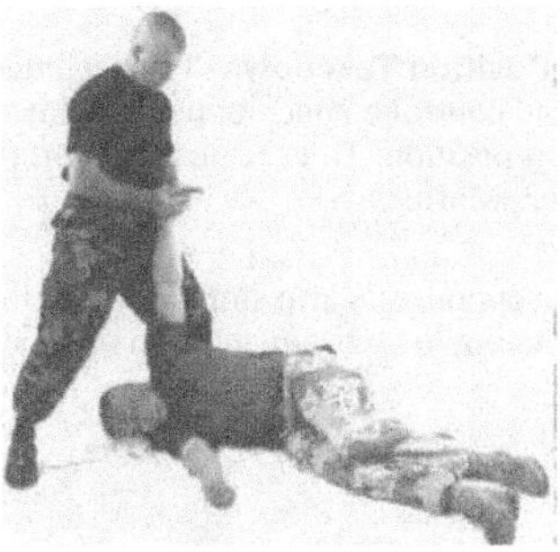

Continue applying pressure with the knee on the opponent's arm. Pivot and step around the opponent's arm to roll him on his stomach.

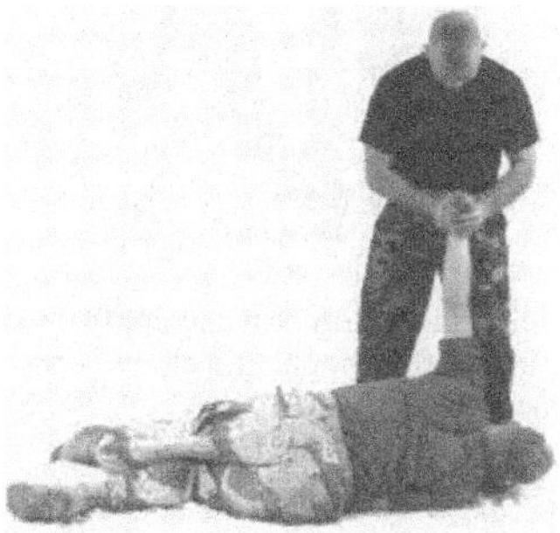

Kneel down with one knee on the opponent's back. The other knee is placed on the opponent's neck and shoulder on either side of his arm. Apply inward pressure with the knees to lock his arm in place.

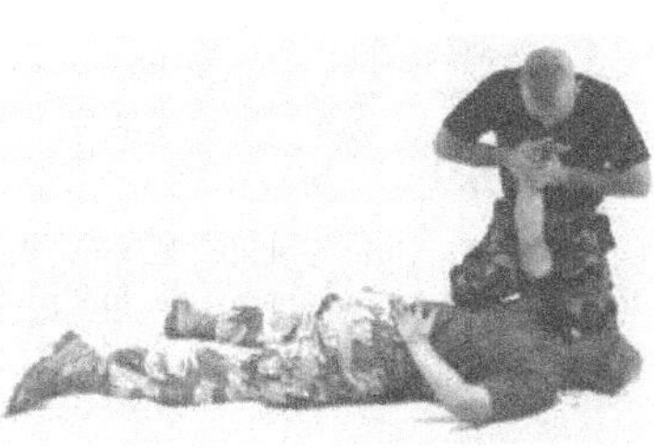

Tell the opponent to put his other hand in the middle of his back. Bring the opponent's controlled hand to the center of his back.

Escort Position Takedown. This technique can be used to control a noncompliant opponent from the escort position. To execute the escort position takedown, Marines—

Lock the opponent's arm straight across the body while rotating his wrist away from the body.

Use the left hand or forearm and apply downward pressure above the opponent's elbow where the triceps meet.

Step back with the right foot and, keeping the opponent's hand controlled against the hip, pivot to the right while continuing to apply downward pressure on his arm to bring him to the ground.

Note: This technique works well when escorting a subject on the Marine's right or left side. When taking down a subject from the right side, step back and pivot to the left.

2. Nonlethal Baton

A baton or nightstick can be an effective compliance tool when used correctly. Batons can be used defensively (blocking), offensively (striking), and as a restraining device when needed. In the fourth level of the continuum of force (assaultive [bodily harm]), defensive tactics include baton or nightstick blocks and blows. Blows to the head or other bony parts of the body are considered deadly force. When deadly force is not authorized, Marines must be able to employ blocks, strikes, and restraints with the baton with the minimum of force.

Grips

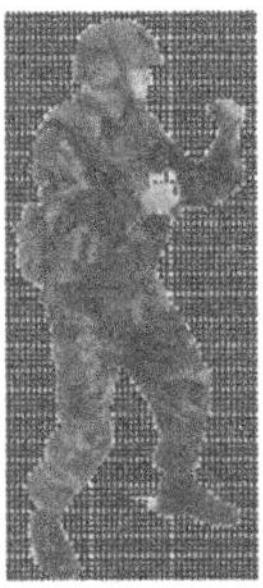

One-Handed Grip. To execute the one-handed grip, Marines—

Use the right hand to grasp the lower end of the baton, about 2 inches from the end.

Wrap the thumb and index finger around the baton so they are touching one another. The grip on the baton should be firm, but natural.

Two-Handed Grip. To execute the two-handed grip, Marines—

Use the right hand to grasp the lower end of the baton, about 2 inches from the end. Wrap the thumb and index finger around the baton so they touch one another.

Use the left hand to grasp the upper end of the baton, palm down, about 2 inches from the end. The hands should be approximately 10 to 12 inches apart.

Stance and Method of Carry

The basic warrior stance serves as the foundation for initiating nonlethal baton techniques. The method of carry provides effective defensive positions with a wide range of options to control a combative opponent.

One-Handed Carry. To execute the one-handed carry, Marines—

Grip the baton using the one-handed grip.

Elevate the baton, with the gripping hand at a level between the belt and shoulder.

Keep the left hand in the position of the basic warrior stance.

Two-Handed Carry. This carry is effective for blocks. To execute the two-handed carry, Marines—

Grip the baton using the two-handed grip.

Elevate the baton, with the left hand higher than the right hand.

Orient the weapon toward the subject.

Movement

In a nonlethal confrontation, movement may be made to create distance between Marines and the opponent or to close the gap to control him. When facing an opponent, Marines move in a 45-degree angle to either side of the opponent. Moving at a 45-degree angle is the best way to both avoid an opponent's strike and to put Marines in the best position to control the opponent.

Blocking Techniques

One-handed blocks are used when carrying the baton in the one-handed carry. The same one-handed blocks used in combative stick techniques (see page 3-6) apply in nonlethal baton engagements. The blocks include blocks for a vertical strike, a forward strike, and a reverse strike. Because the baton is often carried with two hands, there are also two-handed blocks that are effectively used from this carry. Two-handed blocks are discussed in the following subparagraphs.

High Block. Marines execute a high block to deter a downward vertical attack directed at the head and shoulders. To execute the high block, Marines—

Raise the baton up to a level even with or above the head. The baton should be in a horizontal position to block the blow. The fingers of the left hand should be open and behind the baton.

Place the baton perpendicular to the opponent's striking surface to absorb the impact of the blow.

Bend the elbows slightly to help absorb the impact of the blow. The arms should give with the strike of the blow.

Low Block. Marines execute a low block to deter an upward vertical attack directed at the abdomen, groin, or torso. The opponent's blow can be delivered by a foot, knee, or fist. To execute the low block, Marines—

Lower the baton to a level even with or below the groin. The baton should be in a horizontal position to block the blow. The fingers of the left hand should be open and behind the baton.

Place the baton perpendicular to the opponent's striking surface to absorb the impact of the blow.

Bend the elbows slightly to help absorb the impact of the blow. The arms should give with the strike of the blow.

Right Block. Marines execute a right block to deter an attack directed at the head, neck, flank, or hip. The opponent's blow can be delivered by a foot, knee, fist, or elbow. To execute the right block, Marines—

Thrust the baton in a vertical position to the right side.

Pivot to the right by stepping with the left foot and pivoting off the ball of the right foot. Rotate the hips and shoulders into the direction of the block. The fingers of the left hand should be open and behind the baton.

Place the baton perpendicular to the opponent's striking surface to absorb the impact of the blow.

Bend the elbows slightly to help absorb the impact of the blow. The arms should give with the strike of the blow.

Left Block. Marines execute a left block to deter an attack directed at the head, neck, flank, or hip. The opponent's blow can be delivered by a foot, knee, fist, or elbow. To execute the left block, Marines—

Thrust the baton in a vertical position to the left side.

Pivot to the left by stepping with the right foot and pivoting off the ball of the left foot. Rotate the hips and shoulders into the direction of the block. The fingers of the left hand should be open and behind the baton.

Place the baton perpendicular to the opponent's striking surface to absorb the impact of the blow.

Bend the elbows slightly to help absorb the impact of the blow. The arms should give with the strike of the blow.

Middle Block. Marines execute either a left or right block to deter an attack directed at the face, throat, chest, or abdomen. To execute the middle block, Marines—

Thrust the baton in a vertical position straight out in front of the body. The fingers of the left hand should be open and behind the baton.

Place the baton perpendicular to the opponent's striking surface to absorb the impact of the blow. The baton should be held with the left hand forward of the right.

Bend the elbows slightly to help absorb the impact of the blow. The arms should give with the strike of the blow.

Restraining Technique

The strong-side armlock is used to restrain an opponent who is not compliant. To execute the strong-side armlock, Marines—

Use the right hand to run the baton up under the opponent's left armpit, parallel to the ground.

Use the right foot to step forward at a 45-degree angle to the left side of the opponent. The baton is across his forearm.

Use the right hand to drive the baton forward and up so the action bends the opponent's arm behind his back. At the same time, continue moving around him to get behind him.

Place the baton on the opponent's forearm with the thumb and/or fingers and apply pressure to his forearm. At the same time, grasp the other end of the baton with the left hand.

Pull up on the low end of the baton with the right hand. At the same time, push down on the top end of the baton with the left forearm, reaching around with the left hand to grasp the opponent's biceps or shoulder.

Continue exerting downward pressure with the left forearm while pulling back on the opponent's biceps with the left hand. This places the opponent in a position where he is controlled and can be moved.

Apply pressure with the foot against the bend in the opponent's leg above his calf. This lowers the opponent to the ground rather than throwing him to the ground and risking severe injury.

Striking Target Areas

Marines must avoid striking an opponent in the head, neck, or other bony parts with the baton because this is considered deadly force and can lead to serious bodily injury or death. Instead, the legs, arms, and buttocks are target areas that are considered nonlethal.

Legs. Primary targets are the thighs and lower legs. Avoid striking the knee and ankle joints because this can cause permanent damage.

Arms. Primary targets are the upper arms. Avoid striking the shoulder and elbow and wrist joints because this can cause permanent damage.

Buttocks. Primary targets are the buttocks. Avoid striking any other part of the torso, including the chest, rib cage, spine, tail bone, and groin because strikes to these areas can cause permanent damage or death.

One-Handed Striking Techniques

One-Handed Forward Strike. A forward strike follows either a horizontal line or a downward diagonal line using a forehand stroke. To execute the one-handed forward strike, Marines—

Stand facing the opponent with the baton carried in a one-handed carry.

Place the right hand palm up, swing the baton from right to left, and make contact with the opponent.

One-Handed Reverse Strike. A reverse strike follows either a horizontal line or a downward diagonal line using a backhand stroke. To execute the one-handed reverse strike, Marines—

Stand facing the opponent with the baton carried in a one-handed carry.

Bend the right arm and cross the arm to the left side of the body. The baton should be close to or over the left shoulder.

Place the right hand palm down, swing the baton from left to right, and make contact with the opponent.

Two-Handed Striking Techniques

Two-Handed Forward Strike. This strike is an effective follow-up to a middle block or left block. To execute the two-handed forward strike, Marines—

Pull back with the left hand while driving the right hand forward toward the opponent. The baton should be horizontal to the ground. Power is generated by stepping forward with the right foot and rotating the right hip and shoulder into the strike.

Contact the opponent with the end of the baton.

Two-Handed Reverse Strike. This strike is an effective follow-up to a middle block or left block. To execute the two-handed reverse strike, Marines—

Pull back with the right hand while driving the left hand forward toward the opponent. The baton should be horizontal to the ground. Power is generated by stepping forward slightly with the left foot and rotating the left hip and shoulder into the strike.

Contact the subject with the end of the baton.

Front Jab. This strike is effective for countering a frontal attack. It can also be executed as a quick poke to keep a subject away. To execute the front jab, Marines thrust both hands forward in a quick jab. The baton is held either horizontal to the ground or at a slight downward angle.

Rear Jab. This strike is effective for countering a bear hug from the rear. To execute the rear jab, Marines thrust both hands rearward in a quick jab. The baton is held either horizontal to the ground or at a slight downward angle.

APPENDIX A

PUGIL STICK TRAINING

A pugil stick is a training device used to simulate a rifle bayonet so that effective, but safe, training can be conducted to build proficiency of rifle bayonet techniques. Pugil stick training builds on the techniques used to throw punches. Pugil stick training is the only "full contact" training provided to Marines in the Close Combat Program. Pugil stick training teaches Marines to function when faced with stress and violence, and it prepares them to deliver a blow and take a blow. It also provides them with the physical and mental skills vital to success on the battlefield.

1. Pugil Stick Training

Design

A pugil stick consists of a stick wrapped in padding at both ends that can be gripped like a rifle. The pugil stick is approximately the same weight and length of an unloaded rifle with a bayonet attached.

Holding the Pugil Stick

The pugil stick is held in the same manner as the service rifle. All movements come from the basic warrior stance. To hold the pugil stick correctly, Marines—

- Use the right hand to grasp the lower end of the pugil stick overhanded.
- Use the left hand to grasp the upper end of the pugil stick underhanded.
- Use the right forearm to lock the lower end of the pugil stick against the hip.
- Orient the blade end of the pugil stick toward the opponent.

Safety Equipment

The following safety equipment must be worn during any pugil stick bout.

Groin Protection. Groin protection protects the groin from an accidental blow. It should be pulled high around the waist to protect the groin area, with the concave portion against the body.

Flak Jacket. The flak jacket provides protection to the body. It is worn completely fastened.

Neck Roll. The neck roll prevents whiplash if Marines receive a blow to the head. The neck roll further supports the head and protects the neck from blows. The neck roll is worn above the flak jacket and below the helmet. The tied end of the neck roll faces the front.

Helmet. A regulation football helmet protects the face and head. The helmet must fit snugly and the chin straps must be adjusted and snapped.

Mouthpiece. The mouthpiece is worn on the upper teeth to protect the teeth.

Gloves. Marines wear gloves to protect the hands if pugil sticks do not have gloves built-in.

Pugil Stick Screening

Prior to pugil stick training, instructors must ask participating Marines the following questions. If a Marine answers "yes" to any question, he must see the corpsman or a competent medical authority who evaluates him and determines whether or not the he can participate in the training.

- Are you on light duty?
- Are you restricted to running shoes by a corpsman or a doctor?
- Have you fought in a pugil stick bout within the last 7 days?
- Have you received a blow to the head within the last 7 days?
- Have you had a concussion within the last 6 months?
- Have you had dental surgery within the last 24 hours?
- Do you have stitches or staples on your body?
- Have you had a shoulder or head injury within the last 5 years?
- Are you taking a prescription drug?
- Do you have an ear infection or current sinus infection?
- Have you had a broken bone within the last 6 months?

Safety Personnel

The following safety personnel are required to conduct pugil stick training:

- One close combat instructor must officiate the bout.
- One close combat instructor trainer, commissioned officer, or staff noncommissioned officer must be in the training area to serve as range safety officer.
- One corpsman must be in the training area.

Training Area

To prevent injury, Marines train on areas with soft footing (i.e., sand or grass). Training mats are not recommended because feet can stick to the mats, prohibiting movement or causing joint injuries by twisting a knee. Bouts should not take place on a hard surface area; e.g., a flight deck or parking lot. A boxing ring may be used to conduct pugil stick bouts; ring dimensions can vary as long as there is ample room to execute the techniques in the training area.

Second Impact Syndrome

Second Impact Syndrome occurs when a second blow to the head produces a second concussion that occurs within 1 week following a previous concussion (before recovery from the first concussion). Second Impact Syndrome causes rapid brain swelling and can cause death. Therefore, there must be 7 days between pugil stick bouts to reduce the risk of severe injury resulting from Second Impact Syndrome. The 7 day separation between pugil stick bouts significantly reduces the possibility of injury, particularly in someone who may have suffered a brain injury or concussion but shows no symptoms.

Any Marine who experiences headaches or the following symptoms after training must be examined by appropriate medical personnel:

- Blurred vision.
- Ringing in the ears.
- Dilation of the pupils.
- Slurred speech.
- Bleeding from ears or mouth.
- Swelling in head or neck area.
- Any unnatural discoloration in head or neck area.

The Marine should not be allowed to participate in pugil stick training or any other activity where a heavy blow might be sustained for a minimum of 7 days after the headache or other symptoms have subsided.

Safety Measures

The following safety measures must be followed:

- Gear must be worn properly throughout training.
- Contact lenses or glasses will not be worn.
- False teeth will be removed from the mouth.
- Nothing will be worn around the neck except the neck roll.

- Competition among groups of Marines is authorized as long as it does not overshadow training objectives or compromise safety procedures.
- Safety and proper techniques are paramount.
- Safety is more important than competition.

2. General Rules and Regulations Governing Pugil Stick Bouts

Instructors and Support Personnel Requirements

At a minimum, one close combat instructor observes a pugil stick bout. For safety purposes, it is better to have two instructors judging a bout because each instructor can fully observe each of the fighters. The best position for observation is to the right of a fighter. This allows the instructor to see the fighter's facial expression and body movement. The instructor's position must not interfere with the fight.

In addition, one commissioned or staff noncommissioned officer (to serve as range safety officer) and one corpsman will be in the training area.

Prior to the Bout

Prior to the bout, Marines—

- Are paired according to height, weight, and gender.
- Wear the proper safety equipment.

Once Marines have properly donned the safety equipment, they wait for instructions from the close combat instructor. On the command of the close combat instructor, a pair of Marines will enter the ring. Once in the ring, the close combat instructor inspects each Marine for the proper safety equipment.

Note: There should be enough gear so that when two Marines are training, two other Marines can be donning safety equipment for the next bout.

During the Bout

The bout begins when the close combat instructor blows the whistle. All strikes are directed above the waist. Upon hearing a whistle blast, all fighting immediately ceases.

Stopping a Bout

There are two reasons for stopping a bout: delivery of a scoring blow or an unsafe condition. The close combat instructor trainer, close combat instructor, or the range safety officer may stop the bout at any time an unsafe condition is observed.

Scoring Blow. A scoring blow is an offensive technique delivered to a vulnerable area of an opponent with sufficient force and precision to be considered as a disabling or killing blow. Scoring blows are not judged solely on the degree of force with which a blow is delivered, but on the accuracy and techniques employed. A scoring blow is defined as—

- A straight thrust with the blade end of the weapon (red end of the pugil stick) to the opponent's face mask or throat.
- A slash to the side of the opponent's helmet (below the ear) or neck with the red end of the pugil stick.
- A heavy blow to the opponent's head with an authorized technique (i.e., buttstroke, smash) using the butt (black end) of the pugil stick.

When a scoring blow is delivered, the close combat instructor blows the whistle to stop the bout.

Unsafe Condition. The bout will be stopped as soon as an unsafe condition exists. An unsafe condition exists when a Marine is unable to defend himself, loses his balance and falls down on one or two knees or falls down completely, shows instability (e.g., buckling at the knees), loses muscular tension in his neck and his head snaps back or to one side, or appears disoriented. An unsafe condition also exists if a Marine lets go of one end of the pugil stick, the equipment (e.g., helmet, neck roll) falls off, or a Marine fails to use the proper techniques.

If any of these conditions occur—

- The close combat instructor trainer, close combat instructor, range safety officer, or anyone supervising training stops the bout and separates the two Marines.
- The corpsman evaluates the possibly injured Marine to determine if the fight can continue. The corpsman—
  - Ensures that he is alert and responsive, both verbally and physically.
  - Talks to him to see if he is coherent.
  - Makes certain he comprehends and replies to verbal questions.
  - Makes certain his speech is not slurred.
  - Checks his physical signs.
  - Ensures his eyes are focused and not dazed or glazed.
  - Ensures his legs are not wobbly or shaky.

The pugil stick is not used as a baseball bat. The use of unauthorized techniques will result in expulsion from the bout. Only techniques taught in bayonet training are authorized. These techniques include—

- Straight thrust.
- Buttstroke (horizontal and vertical).
- Smash.
- Slash.
- Parry.
- Blocks (low, high, left, right).

After the Bout

A third whistle blast is used prior to identifying the Marine who delivered the killing blow.

A fourth whistle blast is used prior to demonstrating the killing technique used by the Marine who won the bout.

3. Directions for Making Pugil Sticks

Close combat instructor trainers are taught to make pugil sticks to standard specifications at school. It is recommended that all pugil sticks be made under the supervision of a close combat instructor trainer. Until the Marine Corps acquires a standardized pugil stick that units can order through the supply system, instructors use the following information and instructions to make pugil sticks locally.

Supplies

The following supplies are needed to make one pugil stick:

- One 45" by 2" circular, oak, wood dowel.
- Four 9" wide by 27" long foam pad pieces to make two inner handguards of double thickness.
- Four 11" wide by 19" long foam pad pieces to make two outer handguards of double thickness.
- Canvas material to cover inner and outer handguards.
- One foam pad cut 1/4" thick by 7" wide by 13" long to make the centerguard.
- One 8" wide by 14"long piece of canvas material to make a pugil stick sock.
- Canvas material to make two pugil stick end caps. Ideally, two different color fabrics should be used to simulate the bayonet and buttstock ends of a rifle. Common colors for end caps are black and red. Each end cap is a different color to help the officiating certified close combat instructor determine the scoring blow during the pugil stick bout.
- Two 1/4" thick by 11" wide by 54" long foam pad pieces to make two ends.
- Two foam pads cut 1/4" thick by 3" wide by 14 1/2" long to make two end plug inserts.
- One can of 3M Photo Mount™ spray adhesive (NSN 8040-01-120-4009)to adhere foam padding to the stick. (Spray adhesive is available from self-service.)
- One roll of duct tape or riggers tape (NSN 7510-00-266-5016) to protect foam pad and

canvas from wear and to reinforce gluing action of spray adhesive. (Duct tape or riggers tape is available from self-service.)
- One roll para-cord or 5-50 cord to tie both the inner and outer handguards together.
- One roll of thread strong enough to sew canvas material together.
- Sixteen grommets, one for each of the four corners of the inner and outer handguards.
- One foam sleeping mat (NSN 8465-01-109-3369). Use unserviceable foam sleeping mats obtained from Defense Reutilization and Marketing Office.

Tools

The following tools are needed to make a pugil stick:

- Tape measure, ruler, or yardstick.
- Scissors.
- Grommet machine.
- Knife.
- Sewing machine.

Directions

To make a pugil stick, Marines perform the following steps.

Preparing Materials

Step 1. Cut foam sleeping mat into the dimensions listed above. Cut mats to double thickness.

Step 2. Use spray adhesive to adhere the two 9" by 27" pieces of foam together. Use the spray adhesive to adhere the two 11" by 19" pieces of foam together. Let adhesive dry before using the double-thick foam for handguards.

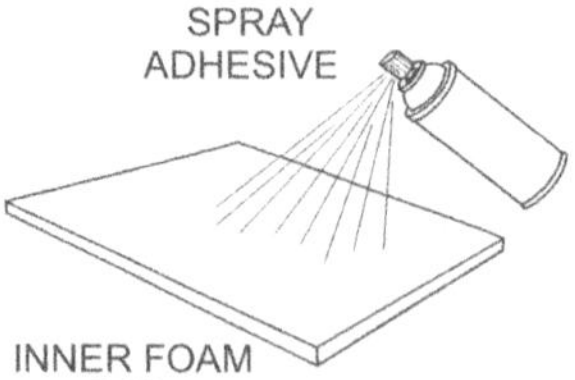

Step 3. Cut canvas material to make two coverings for the inner handguards. Cut material to make two coverings for the outer handguards. Two inner handguards and two outer handguards are required for each pugil stick.

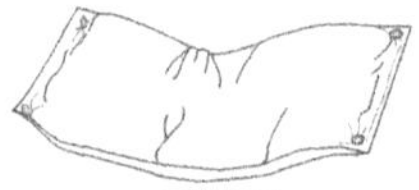

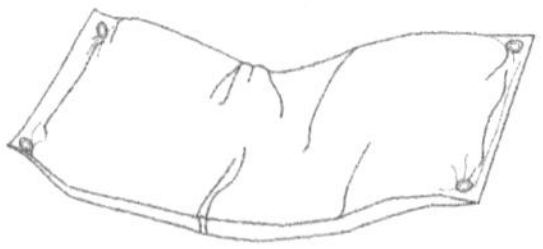

Step 4. Use a sewing machine to make a sleeve out of the 9" by 27" canvas. Sew down three sides. Insert foam pad inside the canvas sleeve and then sew down the fourth side. Use the grommet machine to create holes reinforced with metal grommets in each of the four corners. Once finished, the inner handguards are complete.

Note: It is recommended that the Fabric Repair Shop be used to sew canvas sleeves and apply grommets.

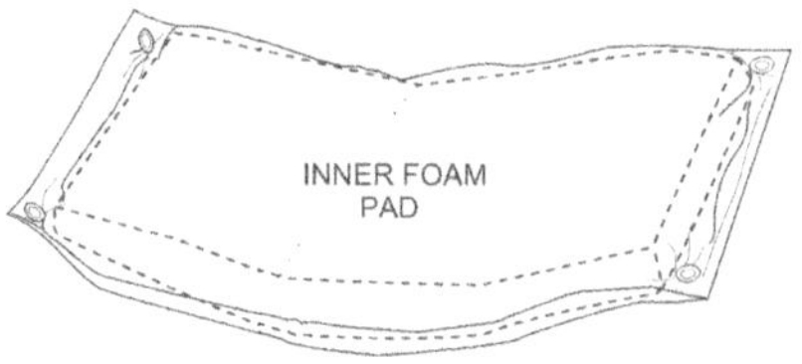

Step 5. Repeat step 4 to make the other inner handguard and both outer handguards.

Constructing Ends, End Plugs, and End Caps

Step 1. Apply spray adhesive on approximately 8 inches of one end of the dowel. This adhesive serves to glue the foam pad to the wooden dowel.

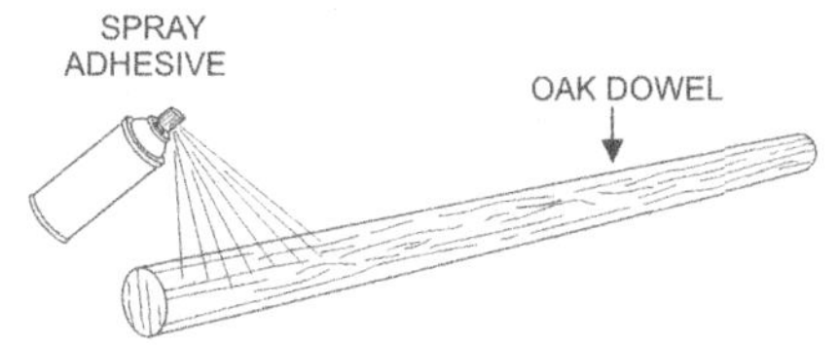

Step 2. Apply spray adhesive on one side of the 11" by 54" foam pad. This causes the foam pad to stick to itself when it is rolled around the end of the wooden dowel. Ensure 3 inches of the rolled foam pad extends past the wooden dowel for stability.

Step 3. Apply spray adhesive on one side of 3" by 14 1/2" foam pad. Roll pad into a tight cylinder.

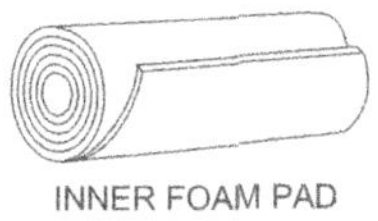

Step 4. Apply spray adhesive on the exposed end of the dowel and fill the hole in the end with the end plug. The end plug prevents the wooden dowel from protruding beyond the edge of the rolled foam padding. This is an important safety precaution to prevent injury.

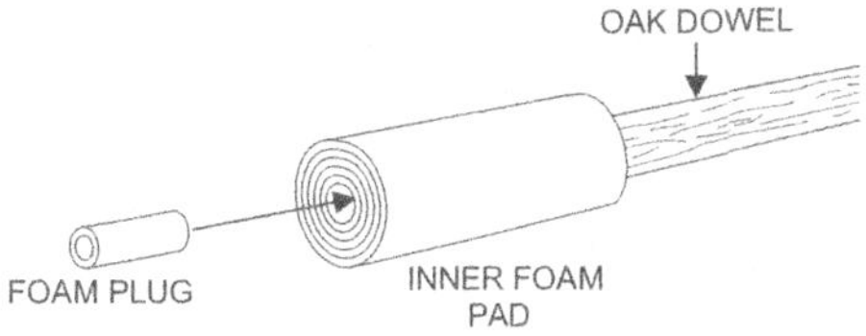

Step 5. Use duct tape or riggers tape to secure and reinforce the foam pad once the end plug is inserted in the end.

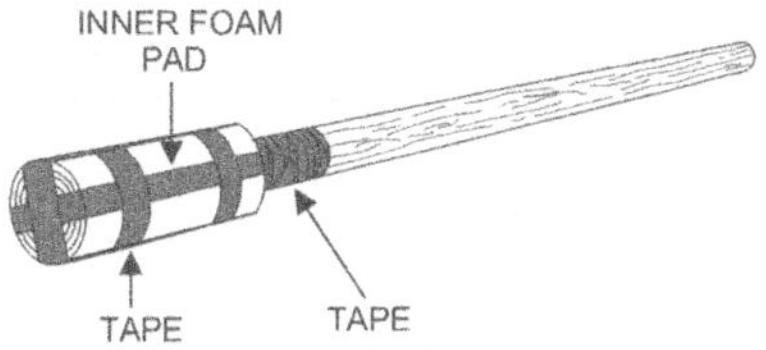

Step 6. Measure the dimensions of the ends (typically, the length of the cylinder should be 12 inches long to allow for taping). Cut and sew canvas material in the shape of a cylinder with one open end. Once complete, this becomes the end cap. The end cap is a colored sleeve that covers the rolled and taped foam pad at each end of the pugil stick.

Note: It is recommended that the Fabric Repair Shop be used to make the end cap.

Step 7. Slide one of the end caps over the rolled foam pad and tape it to the dowel.

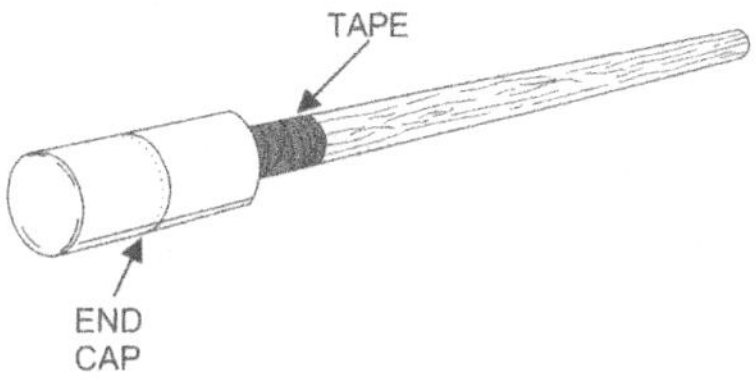

Building the Centerguard

Step 1. Spray one side of the 7" by 13" foam pad with spray adhesive and wrap it tightly around the center of the dowel.

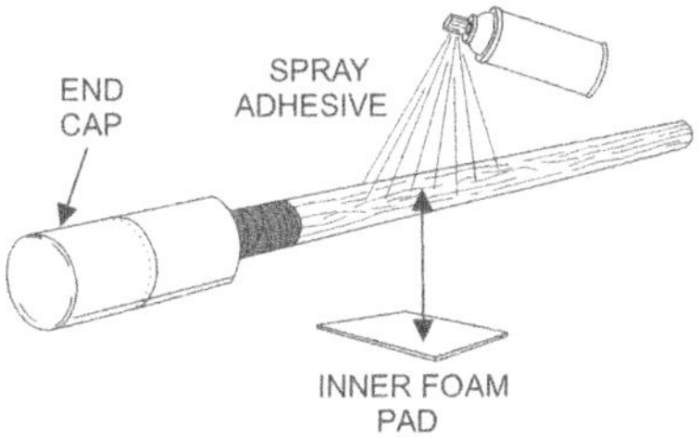

Step 2. Wrap duct tape or riggers tape around the foam pad to secure.

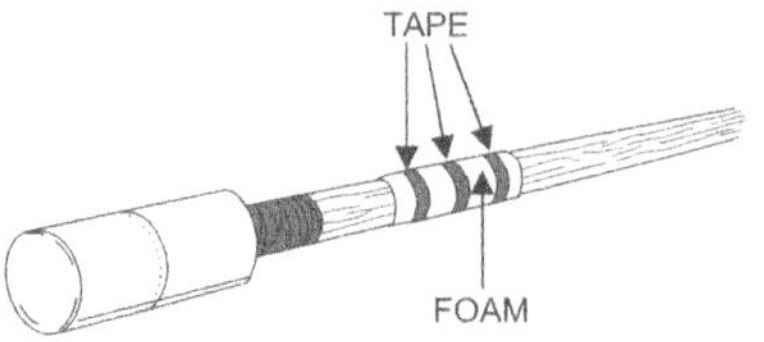

Step 3. Use a sewing machine to make a tubular pugil stick sock out of the 8" by 14" piece of canvas. Slide the pugil stick sock over the rolled foam pad and secure it with the duct tape or riggers tape. Duct tape or riggers tape protects the edges of the centerguard and reinforces the gluing effects of the spray adhesive.

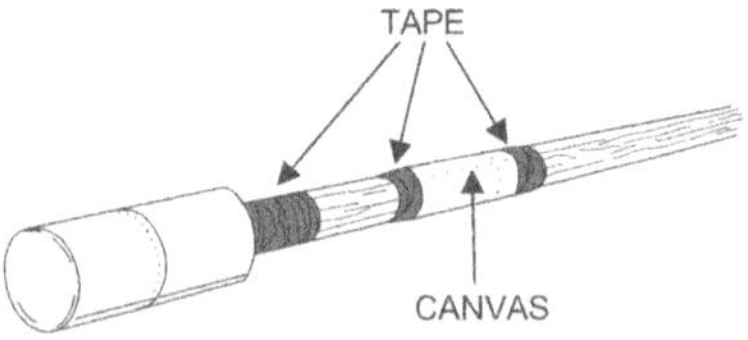

Note: It is recommended that the Fabric Repair Shop be used to make the pugil stick sock.

Finishing the Opposite End Cap

After one end cap is complete and the centerguard is finished, then repeat steps 1 through 7 in Constructing Ends, End Plugs, and End Caps for the other end of the pugil stick. Ensure that the opposite color end cap is used.

Fastening Handguards

Step 1. Place the inner handguards (longer piece) around the dowel. The inner handguard is placed on first because it protects the wrist. One inner handguard is needed for the left hand and one for the right hand.

Step 2. Fold the inner handguard around the pugil stick and run the para-cord or 5-50 cord through the metal grommets.

Step 3. Tighten down the corner ends of the inner handguards by pulling on the para-cord or 5-50 cord.

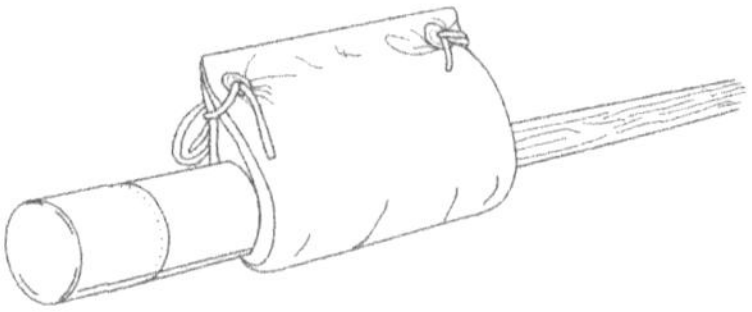

Step 4. Tie off the cord with a square knot above the metal grommets.

Step 5. Place the outer handguards (shorter piece) over the inner handguards (longer piece). Repeat steps 1-4 for the other handguard. The outer handguards provide additional protection to the knuckles and fingers and provides additional stability for the inner handguards.

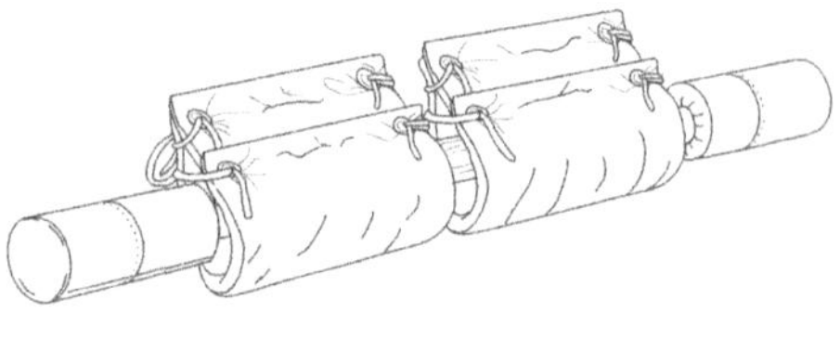

Serviceability Inspection

Prior to any training, all pugil sticks are inspected to ensure they are fit for use. If any of the following conditions exist, the pugil stick is unsafe and will not be used:

- Stick protrudes through the end caps.
- Frayed or worn edges exist on canvas surfaces.
- The stick can be felt through the pads or the padding is too soft.
- End caps or handguards are not securely attached.

(reverse blank)

APPENDIX B

SAFETY PRECAUTIONS DURING TRAINING

1. General Safety Precautions

When training close combat techniques, certain safety precautions must be adhered to to prevent injuries.

Most training should be conducted on a training area with soft footing such as a sandy or grassy area. If training mats are available, they should be used. A hard surface area is not appropriate for close combat training.

All techniques should be executed slowly at first. Marines can increase the speed of execution as they become more proficient. Marine-on-Marine training that requires contact (chokes, throws, ground fighting, and unarmed restraints and manipulation) should not be executed at full force or full speed.

If a technique is applied to the point that a Marine is uncomfortable, the Marine must "tap out." This indicates immediate release of the pressure being applied or to immediately stop the technique. The Marine "taps out" by firmly tapping his hand several times on any part of the opponent's body that will get his attention or by saying *stop*.

Second Impact Syndrome occurs when a second concussion develops within hours, days, or weeks following a prior concussion (and before recovery from the first concussion). Second Impact Syndrome causes rapid brain swelling and can cause death. Marines who experience headaches or other symptoms following training must be examined by appropriate medical personnel. These symptoms can include, but are not limited to, blurred vision, ringing in the ears, dialation of the pupils, bleeding from the ears or mouth, slurred speech, swelling in head or neck area, or any unnatural discoloration of head or neck. They should not be allowed to participate in pugil stick training or any other activity where a heavy blow might be sustained for a minimum of 7 days after the symptoms have subsided.

2. Safety Precautions for Individual Techniques

Falls

When training Marines to fall, they should progress from the ground, to a kneeling or squatting position, and then to a standing position. This ensures they are comfortable and understand the technique before progressing to executing falls from a higher profile. This instructional technique vastly reduces the risk of injury.

Strikes and Punches

When training Marines to strike and punch, they begin by executing the techniques "in the air." As they become more proficient, they execute strikes on equipment (when available) such as an air shield or a heavy bag. At no time should they be permitted to execute strikes on another student.

Chokes and Ground Fighting Chokes

When training Marines to execute chokes, they will not apply pressure to the opponent's throat during training because the trachea and windpipe can be crushed. During training, Marines should practice the proper procedures for blood chokes. They should never execute air chokes. At no time during training should a choke be applied at full force or full speed. No choke should be held for longer than 5 seconds.

Unarmed Restraints and Manipulation

When training Marines to execute unarmed restraints and manipulation techniques, they utilize slow and steady pressure. Never apply these techniques at full force or at full speed.

Pugil Stick Training

A blow to the head during training will bruise the brain. A second blow to the head can cause death. This is known as the Second Impact Syndrome. There must be 7 days between pugil stick training to prevent injury or death.